ROUTLEDGE LIBRARY EDITIONS:
ECONOMIC GEOGRAPHY

Volume 8

T0264541

THE GEOGRAPHY OF IRON
AND STEEL

THE GEOGRAPHY OF IRON AND STEEL

NORMAN J. G. POUNDS

Routledge
Taylor & Francis Group

LONDON AND NEW YORK

First published in 1959
Second revised edition 1963
Third revised edition 1966
Fourth revised edition 1968

This edition first published in 2015
by Routledge
2 Park Square, Milton Park, Abingdon, Oxon, OX14 4RN

and by Routledge
711 Third Avenue, New York, NY 10017

Routledge is an imprint of the Taylor & Francis Group, an informa business

British Library Cataloguing in Publication Data
A catalogue record for this book is available from the British Library

ISBN: 978-1-138-85764-3 (Set)
eISBN: 978-1-315-71580-3 (Set)
ISBN: 978-1-138-86051-3 (Volume 8)
eISBN: 978-1-315-71642-8 (Volume 8)
Pb ISBN: 978-1-138-86054-4 (Volume 8)

Publisher's Note
The publisher has gone to great lengths to ensure the quality of this reprint but
points out that some imperfections in the original copies may be apparent.

Disclaimer
The publisher has made every effort to trace copyright holders and would
welcome correspondence from those they have been unable to trace.

THE GEOGRAPHY OF
IRON AND STEEL

Norman J. G. Pounds

HUTCHINSON UNIVERSITY LIBRARY
LONDON

HUTCHINSON & CO (*Publishers*) LTD
178–202 Great Portland Street, London W1

London Melbourne Sydney
Auckland Bombay Toronto
Johannesburg New York

First published 1959
Second (revised) edition 1963
Third (revised) edition 1966
Fourth (revised) edition 1968

© Norman J. G. Pounds 1959, 1963 and 1966

This book has been set in Times, printed in Great Britain on Smooth Wove paper by Anchor Press, and bound by Wm. Brendon, both of Tiptree, Essex

09 051001 1 (cased)
09 051002 X (paper)

CONTENTS

MAPS AND DIAGRAMS

FOREWORD TO SECOND EDITION

To the list of those who in many ways have helped him in preparing the second edition of this book, the author wishes to add Professor Robert N. Taaffe who has read critically the material on the Soviet Union and made many more suggestions than could, for reasons of space, be included.

N. J. G. P.

Indiana University

FOREWORD TO FIRST EDITION

This is a very small book about a very large subject. Much has been condensed and more has had to be omitted. From among many possible lines of approach, the author has deliberately chosen the historical, seeing in the present pattern of industry a kind of palimpsest, made up of the sequent patterns drawn over a period of several centuries. Each of these geographical patterns springs from the needs, the scientific knowledge and the technology of its own day and age. Each has continued to exist, in some shape or form, after the forces which created it had disappeared. Hence, in this book, there is an emphasis on the history of the technology of iron and steel.

The author wishes to express his thanks to Professor Thomas R. Smith of the University of Kansas for his most critical and helpful reading of the text, and to the following who at some stage read the manuscript, in part or in whole, and made suggestions and corrections:

Professor S. Earl Brown of Ohio State University, Professor W.

Gordon East, Mr Leonard B. Tennyson and his colleagues of the European Coal and Steel Community, Mr James K. Welsh and Mrs Joan Wilson Miller; also to Mr D. R. Baker who drew the maps.

NORMAN J. G. POUNDS

Kansas University
March, 1959

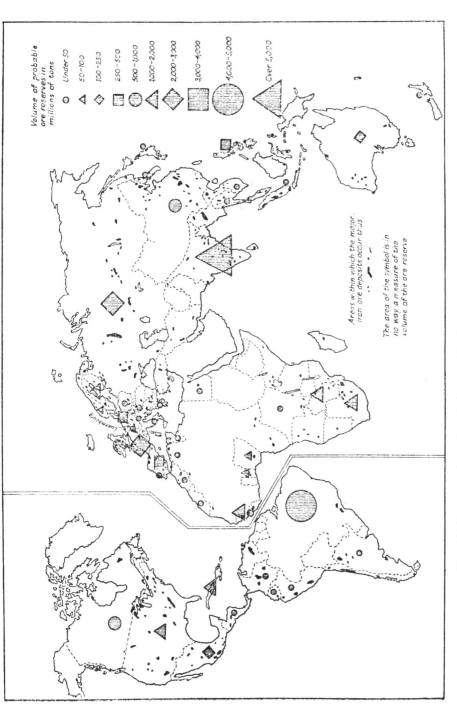

Fig. 1. Iron-ore deposits of the world with amount of probable reserves
(Based primarily on *World Iron Resources and their Utilization*, U.N., 1950, and *Survey of World Iron Ore Resources*, U.N., 1955)

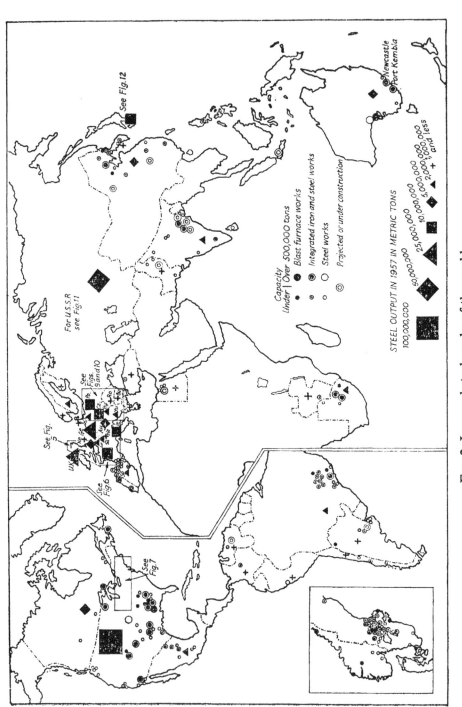

FIG. 2. Iron and steel works of the world
(Based on *Iron and Steel Works of the World*, ed. H. G. Cordero)

1

THE ART AND SCIENCE OF IRON-WORKING

EARLY in the second millennium before Christ the process of smelting iron was discovered. The discovery was made in the Middle East, but by what combination of accident and intuition we do not know. Knowledge of the process spread slowly, first to Egypt and then to the Aegean, where, even in Homeric times, iron was regarded as a rare metal and weapons were made of brightly shining copper.[1] The use of iron reached the basin of the Upper Danube by about 900 B.C., and from this area it was carried by the migrating Celts westwards into France and the Iberian peninsula, and north-westwards across Germany to the British Isles.

Iron-ore is one of the most widely distributed of the metallic minerals. Primitive man must on countless occasions have built his hearth of lumps of iron-bearing rock, and sooner or later he was bound unwittingly to achieve the conditions under which the ore could be reduced to an impure metal. The ore consists of metallic iron in chemical combination with varying quantities of oxygen, carbon and sulphur. Other substances, especially phosphorus and manganese, may also be present in the ore and influence, adversely or otherwise, the quality of the metal that may be obtained from it.

Some constituents of the ore can be driven off by the simple process of heating, but the removal of others can be achieved only by their chemical combination with oxygen or carbon. Yet others have defied the skill and ingenuity of the iron-worker, and continue to plague the modern metallurgist.

Pure iron melts at the temperature of 1537°C., but long before this temperature is reached the metal becomes a pasty mass that can be kneaded and shaped by means of a metal rod. Molten iron is capable of dissolving carbon, which in turn has the effect of lowering the melting point of the iron. Indeed, the melting point of the iron can be brought down to 1130° if it can be made to dissolve sufficient carbon.

Such temperatures as these were unlikely to be achieved by primitive man. In fact, it is improbable that a really fluid iron was regularly created before the invention of the blast-furnace in the later Middle Ages. The possibility cannot be discounted, however, that occasionally, even before this, unusually favourable circumstances may have allowed the iron to fuse completely. Indeed, the evidence, both literary and archaeological, compels one to this belief. But it remains generally true that, before the fifteenth century, iron was obtained in Europe as a pasty lump that could be shaped by hammering, not as a liquid that could be run into a mould.

Early iron-smelting was a simple process.[2] Small pieces of ore were heated on a hearth, resembling the traditional blacksmith's hearth. The temperature was raised by a strong draught. At first the wind was funnelled on to the hearth; then a bellows, worked by the strength of man or animal, or by a wheel turning beneath the weight of falling water, was used. The charcoal fuel supplied heat for the process and, in its combustion, combined with the oxygen in the ore to form CO or CO_2. If this were all, the result should have been a mass of pure metal. In reality the ore contained quantities of foreign matter, which, partially fused, went to form a slag. Furthermore, as the metal itself could not be brought to a fluid state, the particles of slag and of unreduced ore did not separate off from the iron. The result was a metal uneven in texture and quality, with particles of foreign matter scattered through it.

Such was the process. It varied enormously in the details of its execution. As it had been arrived at empirically, the essential stages were associated with unessential details, whose careful execution formed part of the 'magic' of smelting. Thus there were regional methods of making iron. Occasionally a local peculiarity of method was an unconscious adaptation to the quality of the local ore; more often it was a local usage, indefensible in scientific terms.

The iron that emerged from the primitive hearth, or 'bloomery', had rarely absorbed much carbon. Apart from the particles of slag, which were all too abundant, and of ore which had not been reduced, the metal was pure iron. It was soft; it could be bent and fashioned easily, but would not take a sharp, cutting edge. It could be welded, but its strength was small. But occasionally the iron-master could turn out a superior metal. A higher temperature, the absorption by the iron of a small percentage of carbon, and the presence, unknown to the iron-workers, of a trace of manganese in the ore, could result in the creation of a metal resembling steel rather than soft iron. It was still impregnated with slag, but it had the qualities of steel: hardness and elasticity. Some parts of Europe, notably the Siegerland, Styria and Sweden, achieved a high reputation for their steel. Its quality was ascribed to the superior knowledge and skill of its craftsmen, when it could as fairly have been attributed to the quality of the manganiferous iron-ores which they used.

Iron and steel were in earlier times regarded as quite separate and distinct substances. But, just as the medieval alchemist attempted to convert base metals into gold, so the iron-worker tried—with rather greater success—to turn iron into steel. But in his own mind he was only practising a successful kind of alchemy. He was converting one substance into another by methods more magical than scientific. The following passage from a medieval treatise describing the manufacture of a steel file shows the air of magic which surrounded what was really a very simple metallurgical process:

'Burn the horn of an ox in the fire, and scrape it, and mix with it a third part salt, and grind it strongly. Then put the file in the fire, and when it glows sprinkle this preparation over it everywhere, and, some hot coals being applied, you will blow quickly upon the whole, yet so that the tempering may not fall off . . . extinguish it in water.'[3]

Expressed in more technical terms, the process described by Theophilus consisted in adding carbon and in heating it until the iron had absorbed or dissolved enough of the carbon to acquire the characteristics of steel.

Thus from the dawn of the Iron Age until the later Middle Ages iron was made on the hearth or bloomery. Occasionally steel, known as 'natural' steel, resulted, but usually only soft, weldable iron, rich in slag and impurities. Steel, if not made directly from the ore on the hearth, was obtained by the process which came later to be called *cementation*, the adding of carbon to the metal without actually melting it. Throughout these ages, iron was an expensive if not rare metal. In the words of Thorold Rogers, 'relatively speaking, iron was considerably dearer than lead, and frequently nearly as costly as copper, tin and brass'.[4] It was used, of course, for tools, weapons and armour. Often enough, only the coulter of a heavy plough and the tip of the share were of iron. A little was used in the great buildings of classical and medieval times, often in the form of ornamental iron grill-work, for which soft iron was particularly suited. But iron was unknown in the kitchen. The carpenter had generally to work without nails; wire was rare, and a needle was almost considered an heirloom. Yet the manufacture of iron was carried on widely in medieval Europe if not also in the rest of the ancient world. Iron-ore is one of the most widely distributed of minerals, and few regions were entirely without deposits of some kind. Forests were more extensive in earlier times than today, and the supply of charcoal presented no serious problem. Bloomeries could be—and often were—established in any forested region. The introduction of bellows, worked by water-wheel, to increase the draught, greatly influenced the sites of bloomeries. The banks of a swift stream, rather than a windy hilltop, was preferred. Any area that could offer water-power, as well as charcoal and a modest deposit of ore, enjoyed a competitive advantage that medieval man was soon to recognize. Thus the Weald of Kent and Sussex, the deep valleys of Burgundy and the Rhineland and the foothills of the Alps and Pyrenees, became the centres of an iron industry. But there were few areas, however poorly endowed by nature, that did not at least attempt to supply their needs in this scarce but essential commodity.

The Blast-furnace: The most important technical advance in the history of the iron industry was the invention of the blast-furnace.[5] Yet the blast-furnace was not the creation of an inventive genius; it was developed gradually from the bloomery. If the side walls of the bloomery hearth were built higher, the hearth could take more fuel and ore, but the bloom of iron, forming in the bottom of the hearth, could no longer be manipulated by the iron-worker. On the other hand, the smelting was no longer completely exposed to the air; the draught, entering at the bottom of the furnace, was more effective, and the temperature reached in the hearth much higher than on the older and smaller pattern. In this way the metal that was formed reached a temperature at which it began to melt and to absorb carbon from the fuel with which it was surrounded. This, in turn, lowered its melting point, so that fusion was complete, and a mass of iron flowed into the base of the hearth where it cooled into ingot.

This mass of iron differed fundamentally from the earlier bloom. In the first place, as the fusion of the metal had been complete, particles of slag had separated out to form a crust which floated on top of the iron. Secondly, the metal, though relatively free of other substances, had absorbed carbon, which might have made up 4 per cent or even more of its weight. Such iron could be remelted and cast in a mould, but it was hard and brittle, and of no value for these purposes for which iron was used at this time. It had to be refined, in order to rid it of carbon before it could be forged into a weapon, drawn into wire or made into ornamental iron-work.

The high walls of this rudimentary blast-furnace prevented the ingot of iron from being lifted out. Instead, the walls themselves were broken down, the mass of iron removed and the furnace re-built for its next run. Such furnaces, with their many variations in design, were known as *Blau-*, *Wolfs-* and *Stück-öfen.* They were built in Europe from the fifteenth century, if not earlier, and the last of them did not cease to function until the nineteenth. They were waste-ful of fuel and of time, but their operation became traditional in certain areas, where they remained in use long after furnaces of a more efficient design had been introduced elsewhere.

The next phase in the evolution of the blast-furnace was to make its operation continuous. This was done by constructing a hole, or 'notch', through the wall near the base of the furnace. This hole was filled with clay, and could be pierced and opened at intervals. The metal was thus run out into a sand floor and cast into 'pigs', while fresh fuel and ore were charged into the furnace. In this way a furnace could be run, if desired, until its lining was burned away.

The first blast-furnace was built in the fifteenth century. The exact time and place are unknown, though it probably took place in the Rhineland. The invention altered the scale and nature of iron-working. Whereas a bloomery was easily built and readily abandoned, a blast-furnace represented a large capital investment. It was built in the expectation that it would run for many years. It needed a stronger blast than the bloomery hearth, and this in turn necessitated the construction of a larger water-wheel with its accompanying leat and pond. Its consumption of ore and fuel was much greater than that of the bloomery, and its demands could quickly deplete the surrounding area of timber. It was built, therefore, only where there was ore and charcoal in plenty, and this greatly restricted the area where the industry could be carried on. Distinct iron-working regions began to appear: the Weald and Forest of Dean in England, Champagne, Burgundy, Ariège and Dauphiné in France, the Eifel, Siegerland and Harz Mountains in Germany, Central Sweden, and Styria and Carinthia in Austria.

Whereas in earlier times there had been little trade in iron, and most areas had achieved some degree of self-sufficiency, though at a low level, long-distance trade now became important. The 'Osemund' iron of Sweden sold in many parts of North-western Europe, the iron of Siegen and Styria acquired as wide a reputation as the earlier vaunted metal of Damascus or Toledo.

The Direct and Indirect Processes: A further result of the introduction of the blast-furnace was the division of the iron-making process into smelting and refining. In the older method, soft iron was made 'directly' from the ore on the hearth. Much as it needed further refining, the bloom of iron was the end-product. But the pig-iron, drawn from the blast-furnace, was not an end-product. It had to be refined to make soft iron. So the refinery grew up alongside the blast-furnace. Commonly both were established on the same site, but together they often placed too great a burden on the local supply of fuel and strained the local potential in water-power. The long pigs of iron could be transported on horse or by water from blast-furnace to refinery, and there was every advantage in establishing the latter near the markets. In England, pig-iron smelted in the Forest of Dean or along the Welsh Border often found its way to refineries in the Black Country.[6] Swedish pig-iron was refined in the Low Countries and England, and that of the Siegerland along the numerous streams that flowed down to the Rhine and Ruhr.

Smelting with Coke: For two centuries after the appearance of the blast-furnace there was no important change in the technology of iron- and steel-making. During this period the geographical pattern grew more dense but did not otherwise greatly change. But towards the end of this period the stability of the industry began to be threatened by a shortage of charcoal for the furnaces. Centuries of reckless cutting had taken their toll of the forests, and rival consumers, especially the shipbuilders, were insisting on economy in the use of wood.[7] In England the Wealden industry declined, at least in part, for this reason. But for a long while the attempts to use coal instead of charcoal met with no success. The much advertised claims of Dud Dudley to have smelted with coal at his father's Worcestershire furnaces early in the seventeenth century proved to be unfounded, and success was not achieved for another century.

In 1709 Abraham Darby of Coalbrookdale succeeded in smelting iron with coke made from coal mined on the neighbouring Shropshire coal-field.[8] The failure of earlier attempts to smelt with coal or coke is not difficult to explain.[9] Coke burns less readily than charcoal, a coke-burning furnace must be maintained at a higher temperature than one using charcoal, and requires a much stronger blast. Early attempts to use coke had been made in small furnaces designed to burn charcoal, with the result that the furnace went out or yielded only an impure mass of part-smelted metal. It seems that Darby, more by luck than judgment, had used a furnace of above average size, blown by more powerful bellows than were usual.

The process of smelting with coke spread with a quite remarkable slowness. It made little headway in Great Britain until the second half of the century. Attempts to use it on the continent achieved no success for many years, and it was not until 1796 that the first coke furnace was operated successfully outside Great Britain, in the remote Prussian province of Upper Silesia. In France and Belgium coke began to be used for smelting only after the Napoleonic Wars were over, and the first coke-burning furnace in the Ruhr was not built until 1849.

The smallness of the furnaces used helps to explain the failure of coke-smelting in continental Europe. The Upper Silesian experiment seems to have been successful because a furnace, newly built at Gliwice to English specifications, was used. Elsewhere, the use of coke was successful only when a furnace, designed to take coke fuel, was built for the purpose. But a further reason lay in the quality of the metal which the coke-burning blast-furnace produced. Not only did it have a high carbon content, as was to be expected, but it had absorbed other impurities. The volatile constituents of the coal had largely been driven off in the coke-making process, but some sulphur always remained. One is acutely conscious of it even today in the smell of a coke brazier. In the blast-furnace this sulphur might be oxidized and removed in the waste gases, or it might pass into the metal. At the high temperature of the modern blast-furnace, and with careful control over the slag, the sulphur is largely oxidized, but in the lower temperatures of the furnaces of the eighteenth and early nineteenth centuries this did not happen; the sulphur passed into the metal. The presence of sulphur in considerable amounts made the metal difficult to refine, and almost useless for steel manufacture. In consequence the coke-smelted metal fetched only a low price, and iron-masters clung obstinately to the older methods.

The remedy lay in larger blast-furnaces, a more powerful blast, and the use of hot air instead of cold. The higher temperature of combustion that resulted quickly burned away the walls of the furnace, so that linings of expensive refractories became necessary. The earlier blast-furnaces could be built with comparatively little capital investment by the local mason, but to satisfy these new requirements necessitated a capital investment and a degree of engineering skill that were beyond the scope of the 'small man', who had dominated the industry hitherto.

One of the strangest misconceptions of the older iron-masters was the idea that only cold air should be pumped into the blast-furnace. It was not until 1829 that Nielson introduced at the Clyde iron works the practice of heating the air. He used stoves for heating the blast, but eventually the regenerative principle was applied, by which the hot waste gases from the furnace were drawn through flues, which were heated and then used to channel the blast back to the furnace until they had cooled. The temperature of combustion was raised in this way so much that in Scotland coal was used in the furnace without deleterious results. The impurities in the fuel were either combined with slag or oxidized and removed with the waste gases.

By the middle of the nineteenth century the use of charcoal for smelting was in rapid retreat. Coke had become the dominant fuel, with two significant results. The smelting industry gradually deserted the forests, where it had been carried on for a couple of centuries or more, and came to be located on the coal-fields or within easy access of coal. Secondly, the number of works declined, and the scale of each increased.

Advance in Iron-refining: These advances in iron-smelting upset the balance between smelting and refining. A large volume of the pig-iron smelted went into iron castings, which, in the eighteenth and early nineteenth centuries, were of increasing importance; and the rest to the refineries. No significant advance had been made in the technique of refining for a long while. The process was slow, laborious and costly, and some means of speeding it up, and of bringing it more in harmony with the rate of production of pig-iron, was greatly needed. It was thus that the puddling process, patented in England in 1781 by Henry Cort, spread with quite unusual rapidity.

The primary purpose of refining was to oxidize and remove the carbon and as much of the other impurities as possible. Charcoal, which contained no impurities, was used both to supply heat for the process and to react with the carbon. Coal, and even coke, could not be used because they would impart sulphur to the metal. The puddling process, on the other hand, used a reverberatory furnace. The metal to be refined was placed in the shallow pan of the furnace, while the flames from a coal fire were drawn back over it on their way to the flue. High temperatures were raised in the furnace, but direct contact between metal and fuel was avoided. The chemical change was effected by adding iron-ore (Fe_2O_3), which served as an oxydizing agent, to the pig-iron. As the pig-iron was melting down, the furnace-man 'puddled' it with a long iron rabble, introduced through a hole in the side. He worked the metal about, exposing it evenly to the heat and to the chemical action of the iron oxide. As the carbon was slowly dissipated the metal became more viscous, and as the process neared its end the puddler 'balled up' the iron with his rabble into a rounded mass of wrought, or soft, iron, which he lifted from the furnace with tongs.

Cort's patent specified that the ball of iron should then be rolled between steel rollers into bar-iron. Like the soft iron refined on the hearth, the puddled metal still contained some slag. The effect of rolling was to draw this out into thin threads. The rolled metal had a fibrous structure, and was particularly tough. It served well for railway lines, and it served so well for such articles as heavy chains that it continued to be made for this purpose until quite recent years.

Puddling was extravagant of fuel, and in very few instances were puddling furnaces built far from coal-mines. It was attracted to the coal-fields even more than the smelting industry. It was also extravagant of human effort. Puddling has been described as the heaviest form of labour ever regularly undertaken by man. The physical effort required to manipulate the ball of heavy metal, accompanied by the heat, smoke and glare of the furnaces, made a day of puddling indeed a feat of endurance. Attempts to lighten the burden by mechanical puddling devices met with little success. The real alternative was found when Bessemer and Thomas, Martin and Werner Siemens showed how steel could be made directly from pig-iron.

Steel-making: Steel, we have seen, is an iron that has dissolved carbon to the extent of from 0·5 to 1·5 per cent of its weight. The so-called 'natural' steel had been prepared on the hearth where a bloom, or some part of a bloom, of iron was made to dissolve the requisite amount of carbon. Cementation steel was made by embedding bars of soft iron in powdered charcoal and heating them in a furnace for many days. In the course of this treatment the iron absorbed carbon to form 'blister' steel. But the blister steel was no more even in texture than the bar-iron from which it was made. Bundles of 'blister' steel were heated, hammered out, rebundled, reheated and hammered again. This went some way towards evening out its texture, but still there were many imperfections and shortcomings in the steel that was available for European craftsmen.

Steel obtained from the Orient was better but its cost restricted its usefulness. Most familiar of oriental steels was that which was made in India, but known in the West by the name of the emporium at which it was traded—Damascus steel.

About 1740 Benjamin Huntsman, a watchmaker, exasperated by the poor quality of the steel with which he worked, attempted to improve on the work of the steel-makers. His method was to break blister steel into small pieces and to heat them in a tightly sealed crucible made of a refractory material. He achieved a high enough temperature for the steel to melt and the slag to separate out. The crucible was then broken open, and the steel poured out into an ingot mould preparatory to being rolled into bar or wire. Huntsman was particularly fortunate in finding a refractory clay suitable for his needs. He did not patent his process, and hoped by secrecy to secure the advantages he deserved. Knowledge of the process leaked, however, and by the end of the century crucible steel was extensively made in Great Britain. The process did not, however, spread to the continent before the Napoleonic Wars, and there was an import from England of 'Huntsman' steel. This was cut off by the blockade of the continental ports, and high rewards were offered by the Napoleonic government to anyone who could rediscover the secrets of its manufacture. Though there were several claimants it does not seem likely that complete success was achieved before peace returned in 1815.

Amongst the claimants was Krupp. Whether he had produced crucible steel during the years of blockade may be doubted, but there is no question of his success in the years that followed. Krupp perfected the technique of casting large ingots of crucible steel which were then fabricated into the moving parts of machines, into ships' propellers and into armaments.[10] The works which he established at Essen centred in the crucible steel furnace and from the time of their foundation they were known as the *Gussstahlwerke*. Crucible steel was also made in the neighbouring city of Bochum, and in England its chief home was Sheffield, scene of the early Huntsman discoveries.

But crucible steel could be made only in relatively small quantities and its cost was particularly high. What was wanted was a method of making steel on a massive scale, and this is what Sir Henry Bessemer and his successors provided.

Converter and Open-hearth: In the middle years of the nineteenth century the output of the blast-furnaces went in part into iron-castings, in part to the puddling furnace. Puddled iron was, for the greater part, rolled into sheets, strips and bars, and the most important single item in the long list of iron products was railway lines. Only a small part of the puddled iron went to the cementation and crucible steel works.[11]

The superiority of steel to puddled iron was obvious, especially for the manufacture of such items as railway lines. The cost alone prevented its use. How to make steel as cheaply and on as large a scale as puddled iron was the problem which Bessemer took up. His experiments convinced him that 'atmospheric air alone was capable of wholly decarburizing grey pig-iron, and converting it into malleable iron without puddling or any other manipulation'.[12] The result was the converter, an egg-shaped steel vessel lined with refractory material, into which molten pig-iron was poured. Air was then blown through the liquid from below. The carbon and other impurities were oxidized and removed either as gas or as slag, and the heat generated by the chemical changes served to keep the metal in a molten condition until the process was complete. In the words of Bessemer himself:

> 'In one compact mass we had as much metal as could be produced by two puddlers and their two assistants, working arduously for hours with an expenditure of much fuel. We had obtained a pure homogeneous ten-inch ingot as the result of thirty minutes' blowing, wholly unaccompanied by skilled labour or the employment of fuel; while the outcome of the puddler's labour would have been ten or a dozen impure, shapeless puddleballs, saturated with scoria and other impurities, and withal so feebly coherent, as to be utterly incapable of being rendered by any known means, as cohesive as the metal that had risen from the mould.'[13]

Bessemer published his discovery in 1856. The acclaim with which it was received was suddenly stilled when it was found that comparatively few varieties of pig-iron were capable of being refined in this way. Entirely by accident Bessemer had used a pig-iron that was almost completely free of phosphorus. Those who adopted his process used, more often than not, an iron smelted from phosphorus-bearing ores. In Bessemer's converter the phosphorus remained in the metal, making it brittle—'cold short'—and almost useless. The value of the process was greatly reduced, though it was used wherever a pig-iron could be obtained that was wholly free from phosphorus.

The economies of the Bessemer process were undoubted, and only the phosphorus problem prevented its widespread adoption. For twenty years the best minds in the field of metallurgy gave themselves to it, but it was not until 1878 that the remedy was found.

Before this, however, another refining process had been invented and its use popularized. Pierre Martin, who operated a small iron works at Sireul, near Angoulême, in France, used a furnace resembling somewhat the puddling hearth, but heated by incandescent gases which were burned in it. In this way the pig-iron was melted down. The carbon was removed by the addition of iron oxide. The puddling furnace had not reached a particularly high temperature, and, at the end of the process, the iron was not a liquid but a pasty lump, from which the slag had not been entirely separated. Martin, by maintaining a higher temperature, hoped to keep the metal liquid, and thus to separate the slag completely, until the process was complete. He expected then to pour off the metal into an ingot mould.

The gases, burning in contact with the metal, certainly raised a higher temperature, but without the regenerative furnace, recently invented by Siemens, Martin's process would have been a failure. The regenerative principle has already been referred to (page 18) in connection with the pre-heating of the air admitted to the blast-furnace. Siemens' furnace consisted, in its essentials, of two chambers, each with a labyrinth of brick checkerwork. The products of combustion came out from the furnace through one, while the unburned gases entered by the other. At intervals this movement was reversed, so that the ingoing gases were always receiving from the checkerwork the heat generated by earlier combustion. Only in this way did Martin achieve a temperature that would allow him to pour off a fluid metal.

These two processes, the converter and the open-hearth, invented within ten years of each other, have ever since been in some sense rivals of one another. Both speeded up and increased the scale of steel-making; both, at least in their earlier years, could use only phosphorus-free iron. But there were sharp differences. The converter acted so quickly that precise control over the chemical changes taking place in it was impossible. The open-hearth, on the other hand, operated slowly enough for samples to be taken and the bath of metal to be examined at intervals. Furthermore, the stream of air passing through the converter not only left minute air cavities but also oxidized particles of metal. The addition of *spiegeleisen*[14] to the molten metal at the end of the 'blow' served in some degree to neutralize these effects, but the quality of converter metal was always lower than that of open-hearth.

The open-hearth, which burned a gaseous fuel, was clearly more expensive to operate than the converter, which was able to generate its own heat from its internal chemical changes. For many purposes the price differential between the two was fully compensated by the differences in the quality of the metal. But the open-hearth had one great advantage, which was certainly not realized by its inventor. It could take and refine not only pig-iron, molten or solid, but also scrap-iron and scrap-steel. The waste from the rolling shops and engineering works came to be a major source of raw material for the open-hearth. Indeed, in so far as scrap-steel had already been once refined, its use speeded up the open-hearth process. Scrap, on the other hand, could not be introduced into the Bessemer converter, without lowering the temperature and impeding or stopping the process. The significance of steel scrap, and its influence on the steel-making process used, are discussed later.

Meanwhile, the overriding problem in both processes was that they could not use a pig-iron containing more than a trace of phosphorus. Sir Henry Bessemer has described his searches for a phosphorus-free iron-ore, and his realization that such ores were not abundant.[15] The problem of phosphorus was in fact solved not by the professional chemists and metallurgists employed by the steel works, but by an amateur in the field who earned his living as clerk to the magistrates of a London borough.

The problem was not difficult to state. The phosphorus had to be combined with the slag, and thus removed from the metal. As its chemical reaction was acid it would not react with the siliceous slag that formed in the converter. The solution would normally be to make the slag alkaline in its reaction by the addition of lime. The phosphorus would then be oxidized to phosphorous pentoxide by the flow of air through the metal, and this would combine with lime and pass into the slag.[16] But an alkaline slag would, at the high temperature at which the process operated, react upon and destroy the silica brick, with which the converter was lined. Sidney Gilchrist Thomas, with the help of his cousin, Percy Gilchrist, produced a refractory brick with an alkaline reaction with which to line the converter, by using crushed dolomite. He carried out his experiments mostly at Blainavon in South Wales, but the 'blow' that finally established the success of the process was at Bolckow and Vaughan's works at Middlesbrough in 1879.

The Thomas or basic process spread rapidly, so urgent was the need for it, and within a year of its publication it was in successful use in the Ruhr, in Lorraine, and in Upper Silesia and Moravia. Within a short time the basic lining was fitted to the open-hearth. Although there continued to be some demand for 'acid' or Bessemer steel, the output of 'basic' increased rapidly and the previous heavy dependence on phosphorus-free ores ceased.

At present the efficiency of these older methods of steel-making is being increased by the use of oxygen in both the converter and open-hearth. It raises the temperature of the reaction; it even allows scrap to be charged into the converter, and cuts down on the impurities, like nitrogen, in 'basic' steel.

Only one other method of steel-making has since gained wide acceptance—the electric furnace. The Bessemer process derived its heat from the combustion of the silica, carbon and other impurities in the iron; the open-hearth, from the burning of gases introduced into the furnace. But the electric furnace was heated either by an electric arc drawn between carbon electrodes and the metal or by the resistance of the metal to an induced current. In either case, no fuel material need be introduced into the furnace, and the nature of the slag—either oxidizing or reducing—and the temperature could be controlled with a degree of fineness unattainable by any of the older methods. Furthermore, alloying metals could be introduced more easily than in the open-hearth. But, owing to its heavy demands on power, the electric furnace is expensive to operate. While it is capable of making steel from pig-iron, it is normally used for the further refining of already refined metal. It is charged not with pig-iron from the blast-furnace, like the converter and open-hearth, but with steel scrap from which sulphur, phosphorus and unwanted carbon have already in large measure been eliminated. The world production of electrically refined steel is only a small fraction of the total steel, but it includes steel for ball-bearings, turbine blades and high-speed cutting tools.

Heat-economy: The chemical changes necessary to the smelting and refining of iron take place, as we have seen, only at high temperatures. The supply of heat, in all except the Bessemer process, is one of the principal items of cost in the iron and steel works. On the other hand, heat is produced and wasted at so many points that attempts have been made to conserve it and thus to reduce costs. The use of the regenerative principle in pre-heating the blast in the blast-furnace and of the gas intake of the open-hearth are examples. But much more serious is the loss of heat if pig-iron, run from the blast-furnace, is allowed to cool down, only to be reheated either in the open-hearth or before being placed in the converter. Ideally, the molten metal should be taken from one direct to the other. This, however, is impossible to organize. Instead, a reservoir of molten iron is held in a huge, brick-lined steel vessel, known as a 'mixer'. Little heat need be applied to keep this metal in a liquid state and ready for pouring into 'thimbles' which carry it to the steel works.

After refining the steel is normally poured into ingot moulds, where it solidifies. The mould is 'stripped' from the ingot while the latter is still near white heat. It is placed in a 'soaking' furnace, which runs at a much lower temperature than either blast- or steel-furnace. Here the ingot is maintained at white heat until the rolling-mills are ready to take it. As the ingot passes through the sequence of rolls, which reduce it to strip, sheet, bars, rods or shapes, it loses heat very rapidly, and unless the rolling is completed quickly the steel has to be reheated. The older rolling-mills required that the steel should be passed and re-passed through the same rolls. This necessarily slowed the process so that the metal had cooled before the process was complete. The modern process of continuous rolling, whereby the steel moves continuously onwards through carefully graded rolls without stopping or reversing, secures that the rolling is completed in at most two heats.

The effect of these developments has been to concentrate the steel-making processes at the blast-furnace site and the rolling at the steel mills. It has also been to reduce the need for coal and thus the advantages of a site near a coal-field. Most works constructed within the past forty years are integrated in this way. In some of them the integration goes yet farther. The coke for the blast-furnace is made on the same site. The coke-ovens may themselves be fired, at least in part, with the exhaust gases of the blast-furnaces. These contain combustible carbon monoxide but usually have too much sulphur for them to be used in the open-hearth, mixer or soaking pits. The gas given off in the coke-ovens is used in a variety of ways. It is burned in the open-hearths; it is a source of supplementary heat for the hot blast, and it is burned in the mixer and soaking pit, and may be used to generate electric power for the rolling-mills. The presence of coke-ovens usually necessitates plant for handling coal-tar and at least its more elementary derivatives. An integrated iron and steel works is thus a very large and exceedingly complex unit in the organization of which the need to preserve and use in one process the waste heat and materials of another plays a very significant role.

[1] For the spread of iron-working *see* M. France-Lanord, 'Evolution de la technique du fer en Europe occidentale de la prehistoire au Haut Moyen-Age', in *Le Fer à travers les Ages*, Nancy, 1956, 27–43.
The standard work on the history of the technology of iron and steel in Ludwig Beck, *Geschichte des Eisens*, five volumes, Braunschweig, 1884–1903. Otto Johannsen, *Geschischte des Eisens*, 3rd edition, Dusseldorf, 1955, presents a shorter study. On iron-working in ancient times, *see* R. J. Forbes, *Metallurgy in Antiquity*, Leyden, 1950, and for a brief treatment *A History of Technology*, ed. Charles Singer, E. J. Holmyard and A. R. Bull, volume I, 1954, 592–9, and Volume II, 1956, 55–61. The papers published in *Le Fer à travers les Ages*, Annales de l'Est, Nancy, 1956, contain a great deal of material on early technology. For Great Britain, see H. R. Schubert, *History of the British Iron and Steel Industry*, London, 1957.

[3] *An Essay Upon Various Arts by Theophilus*, trans. Robert Hendrie, Book III, London, 1847, 223. *See also* A. R. J. P. Ubbelohde, 'The Beginnings of the Change from Craft Mystery to Science as a basis for Technology', *A History of Technology*, IV, 1958, 663–681.

[4] J. E. Thorold Rogers, *Six Centuries of Work and Wages*, London, 1884, I, 87.

[5] *See* L. Beck and O. Johannsen, *op. cit.*   [6] *See* below, page 75.

[7] Robert G. Albion, *Forests and Sea Power*, Harvard University Press, 1926; P. W. Bamford, *Forests and French Sea Power*, Oxford, 1956.

[8] Arthur Raistrick, *Dynasty of Iron Founders*, London, 1953.

[9] T. S. Ashton, *Iron and Steel in the Industrial Revolution*, Manchester, 1951, 32–38.

[10] N. J. G. Pounds, *The Ruhr*, London, 1952, 74–76.

[11] *Sir Henry Bessemer, F.R.S., an Autobiography*, London, 1905, 138.

[12] *Ibid.*, 141.   [13] *Ibid.*, 153.

[14] *Spiegeleisen* is an alloy of iron with from 10 to 25 per cent manganese; it serves also to reduce the sulphur in the steel. Ferromanganese has a higher proportion of manganese, but plays the same role in steel-making.

[15] *Op. cit.*, 178.

[16] The formula is: $4P + 5O_2 + 6CaO = 2Ca_3(PO_4)2$. The calcium phosphate thus formed is the basic slag of agriculture.

2

THE ORES OF IRON

ABOUT 5·06 per cent by weight of the earth's crust is composed of iron, which is thus one of the more abundant constituents of the rocks beneath us. But pure metallic iron occurs only very rarely. Minute particles or even small lumps occur in some of the igneous rocks, but the most common natural source of pure iron is the meteorites that have fallen on to the earth's surface from outer space. The heat generated by their rapid passage through the atmosphere has reduced the iron compounds and burned away most other constituents of the original meteorite. For practical purposes, however, we may say that iron does not occur free in nature.

The Nature of the Ores: Iron occurs most often in combination with oxygen, sulphur and carbon. These substances may further be hydrated, that is, combined chemically with water. The iron-ores of industrial and commercial significance today belong to the following groups:

Haematite, sometimes known as specular or kidney iron, is a simple oxide of iron, Fe_2O_3. It occurs both in a crystalline and in an amorphous or powdery form. Its colour varies from black to red. The amorphous form is usually brick-red in colour, and it is this shade that we most frequently associate with iron mines and iron works. The pure ore yields on smelting about 70 per cent of its weight in metal.

Magnetite is a richer but rarer ore. It is also an oxide of iron, but its formula, Fe_3O_4, shows a higher proportion of iron. The pure ore yields 72·4 per cent of its weight in metal, and is the richest of known ores of iron. It is a crystalline substance, black in colour, and, as its name suggests, is strongly magnetic.

Limonite, or brown iron-ore, is an hydrated form of haematite, with the formula $2Fe_2O_3.3H_2O$. The metal content of the pure ore is lower, about 60 per cent, than in those mentioned. It is normally brown in colour, and has either a fibrous or an earthy structure. Limonite forms from the weathering or degeneration of other substances containing iron. It is common in tropical soils, at the outcrops of mineral lodes, and as concretions of bog-ore over the bottom of swamps and marshes, where it results from bacteriological action. The ochres, used in paint manufacture, are forms of limonite.

Siderite, also known as chalybite and spathic iron, is a carbonate of iron, $FeCO_3$ with 48·3 per cent metal. Its colour ranges from yellow through red and brown to black. It normally occurs in close association with impurities which greatly reduce its value.

Other ores of iron are of much less importance. Iron pyrites (FeS_2) is a very common mineral but has no significance for the iron industry because it has been found impossible under commercial conditions to separate the metal from the sulphur. It is, however, an important basis for the manufacture of sulphuric acid. Other ores of iron occur in such small masses, or are so combined with harmful substances such as phosphorus or sulphur, that there is no likelihood of their profitable exploitation.

The Occurrence of the Ores: Iron-ore occurs in a variety of ways, from rounded nodules at the bottom of lakes to masses of great size and hardness penetrating deep into the earth's crust.[1] The origins of iron ore are as varied as the forms they assume. Numerous as they are, they may nevertheless be divided into two major groups: those formed by the solidification of magmas containing iron that have been intruded into the earth's crust, and those deposited from solution on the beds of swamps, lakes and the sea

The ores of igneous origin may be regarded as the primary source of the metal. The magmas emanate from varying depths beneath the surface and vary greatly in mineral composition. Some iron occurs in most magmas, but more often than not it occurs in such very small quantities or is scattered so thinly through the mass as a whole that the rock is not a significant source of the metal. Sometimes the iron minerals segregated themselves from the rest of the magma while the latter was still liquid, to form concentrations of ore large enough to be mined economically. Much of the magnetite ore and some of the haematite was formed in this way, such as the great magnetic deposits at Kiruna in northern Sweden.

Iron-ores may also be formed beyond the actual limits of the intrusive magma. Chemical changes, brought about by the hot gases which emanate from the magma itself, have led to the formation of iron minerals in the surrounding rock.

But most of the ore deposits that are significant in the world today were not formed directly by magmatic segregation, but are the result of secondary processes. Throughout geological time iron-bearing rocks have undergone the processes of erosion, and iron minerals, whether derived from the large concentrations or from thinly scattered particles, have been carried towards the sea. Iron in its *ferrous* (FeO) form is readily soluble in acidulated water, and in this form makes its way to the lakes and seas. There chemical or bacteriological action can bring about its oxidation and deposition as the insoluble *ferric* iron (Fe_2O_3). The presence of decaying vegetation, as in a swamp, restricts the process of oxidation and leads more often to the deposition of iron carbonate. Such conditions existed *par excellence* in the swamps of the Carboniferous period, when the coal measures accumulated. Iron-ores are sometimes found interbedded with the coal, and in the Ruhr and parts of Great Britain these so-called 'blackband' ores have had an important influence on the growth of the iron industry. Sometimes the ores formed during the periods of sedimentation which occurred in the intervals between the periods of accumulation of coal. Such deposits are known as 'clayband' or 'clay ironstone', and are also impure carbonates of iron.

Rounded nodules of iron-ore—usually limonite—are forming in most lakes and swamps today. Their formation is quite rapid, and, though no longer of serious importance, they once played a very significant role in the development of the industry. An illustration in the *Description des Arts et des Métiers* of the 1760's shows 'miners' scraping these nodules of bog-ore from the bed of a lake with implements that resemble shrimping nets.[2]

The ore deposits that accumulated in lakes and swamps are generally small in extent. More often the iron has been precipitated as ferric oxide on the floor of shallow seas, where natural sedimentation was slow. The oxide is usually found to be associated with iron carbonate and also with silica, and phosphorus is usually present in quantities large enough to be harmful to the ore. The texture of such ores varies. Frequently it is oolotic, as in the 'scarpland' ores of England, the '*minette*' of Lorraine and Luxembourg, and the vast Clinton bed of Alabama.

Lastly, iron-ore deposits can sometimes be formed by the chemical breakdown and removal in solution of the other constituents of iron-bearing rocks, leaving an insoluble mass of iron oxide. Many of the red, lateritic soils of the tropics have this origin, and are comparatively rich in iron. Sometimes the iron has been sufficiently concentrated by these means to give the rock an economic value.

The processes that have here been outlined—precipitation of iron oxide and its concentration by chemical and physical processes —have been going on since the earliest geological times. The beds of ore that have been formed have been hardened, folded and eroded in their turn. It is often difficult or even impossible today to suggest the way in which an ore deposit actually originated. Many of the most famous ore-bodies are highly folded; some of sedimentary or residual origin have been metamorphosed into rocks of exceptional hardness, such as the taconite of Minnesota; others, like those of the English Midlands, can be scooped out with an excavating machine. The Wabana ores of Newfoundland are very old oolitic deposits; those of the Midlands, relatively young. Those of Bilbao in Spain, of the Mesabi Range in Minnesota and of Krivoi Rog in the Soviet Ukraine are all probably fossilized residual deposits of great geological age.

Physical and Chemical Properties of Ores: The age, texture and chemical composition of iron-ores are reflected in the difficulties of mining and smelting. Some have a granular or earthy structure, and can be moved easily with excavating equipment. Though the mining of such ores is relatively cheap, their friable nature is not an unmixed advantage. If used untreated in the blast-furnace they would tend either to clog the blast or to blow out at the top with the waste gases. Sintering (*see* page 34) is the usual method of correcting this, but even so the excessively powdery nature of such ores as those of Krivoi Rog in the Ukraine present important technical problems.

By contrast, 'hard' ores are broken up and removed with difficulty. Resort has usually to be had to blasting. Unless crushed such ores smelt slowly in the furnace and reduce its efficiency.

The mode of occurrence of an ore greatly influences its

profitability. Bedded ores, lying at only a shallow depth below the earth's surface, are ideal. They can be worked in open pits by large-scale equipment, and the hazards and high costs of underground mining are avoided. The East Midland ores in England and the similar deposits of Lorraine and Luxembourg; the Erzberg deposits of Styria and part at least of the Lake Superior ores, are of this kind. But such open-cast mining is rarely possible over large areas. Most of the important iron-ore deposits dip, either gently or steeply, away from the surface. Thus, most of the Lorraine ore is now obtained from underground workings, reached by shafts. And as mining is extended it becomes deeper. The Magnitogorsk deposit of the Southern Urals was, when first worked, a hill of ore, but, as the hill has been worked away, it has given place to a pit of growing size.

This, of course, emphasizes the fact that iron-ores are exhaustible and that, sooner or later, the best of them will be finished. Most deposits are relatively small, and it is very easy (page 52) to enumerate the really large deposits of the world. In the early days of iron-working reliance was chiefly placed on small deposits. They often occurred close to the surface and could be reached easily. A furnace was built as close to the deposit as conditions of fuel-supply, water-power and transport allowed, and in many instances it continued to work until the ore deposit was exhausted. As, however, the smelting units grew in size and complexity it was assumed that they would continue to operate until they had become obsolete or worn out. The deposit of ore had to be large enough to supply the furnaces during their expected lease of life. A small deposit could clearly not do this, and the chances of its ever supporting a smelting industry diminished as the industrial plant grew larger. The same process has operated with regard to mining plant. This usually represents a smaller fixed investment than a smelting works, but its cost has nevertheless to be amortized during the life of the deposit. This means, of course, that as long as large-scale equipment is used, small deposits will not be worked and medium-sized will not attract smelting and steel works to their vicinity. This question of the expected life of ore deposits in relation to the length of the working life of mining equipment and of blast-furnace works has become a matter of great importance in the locating of new works, more especially in the Soviet Union (*see* page 157).

The question whether a smelting works should be located in the vicinity of an ore deposit, close to the source of fuel or at an intermediate site, depends in part on the nature of the ore itself. The figures for the metal content of the different ores of iron, given at the beginning of this chapter, are theoretical only. They represent the proportion by weight of metal in pure ores only. But no ores are pure. All contain earthy matter, known as gangue, in greater or smaller quantities, as well as traces of phosphorus, sulphur, titanium, silicon, manganese and other substances. These latter impart certain qualities, desirable or otherwise, to the ore.

The grade of an ore is the percentage of metallic iron that occurs in the ore as it is mined. The magnetite of northern Sweden, with theoretically an iron content of 72·4 per cent, rarely has in fact much more than 60 per cent metal. This is a relatively rich ore. The *minette* of Lorraine has rarely much more than 30 per cent of its weight in iron, and the proportion is often a great deal less. The following table gives the range of iron content in a number of well-known and much-used ore deposits:[3]

England	..	Northamptonshire Jurassic ore	..	35 per cent
France	..	Lorraine *minette*	28–32	,, ,,
Sweden	..	Kiruna magnetite	65–68	,, ,,
U.S.A.	..	Mesabi haematite	51	,, ,,
,,	..	Alabama red ore	31–39	,, ,,
Venezuela	..	Cerro Bolivar haematite	63	,, ,,
Brazil	..	Minas Gerais ,, ...	50–66	,, ,,
U.S.S.R.	..	Krivoi Rog ,, ...	63	,, ,,
,,	..	Magnitnaya Gora magnetite ..	56	,, ,,
Liberia	..	Magnetite	57	,, ,,

Ores with less than 30 per cent iron are low grade. It is not difficult to understand that when such ores are hauled from the mine to the furnace at least 70 per cent of the load must find its way to the slag pile, after having consumed its share of fuel in the blast-furnace. Ore is bulky and heavy and the transport of such ores, at least over long distances, is clearly something to avoid if this can possibly be done. As a general rule it may be said that low-grade ores attract the smelting plant to them, while rich ores—those with 60 per cent or more of metal—can be transported more economically over great

B

distances. How this works itself out in the location of industrial works will be seen in the next chapter.

There is, locally at least, a shortage of iron-ore in the world today, and it is certain that future generations will face real scarcity. High-grade ores are nearing exhaustion with almost frightening rapidity, but there is in reality an abundance of low- and very low-grade ores. Any discussion of ore shortages must be placed in its economic context: how low in the scale of ore grades can we go? A low-grade ore is expensive—relative to its metal content—both to transport and to smelt. The question then arises, how far can the grade of the ore be improved at the mine? In recent years important advances have been made in this technique of *beneficiation*, as the process is called, which increases the percentage of iron in the ore.[4] The ores may be washed, to separate off the lighter gangue materials. Magnets are used to separate off certain iron-bearing minerals, and ores, crushed to a fine powder, are subjected to gravity and flotation processes to remove the lighter gangue. In the United States a steadily increasing proportion of the iron-ores are beneficiated, and at the present time about a third of the ores mined are improved in these ways. This fraction will certainly increase. Attention has been focused in recent years on the very abundant, but low-grade, ore known as taconite, which occurs in northern Minnesota. As it is mined, only 25 to 30 per cent of it is iron. It is a hard, cherty rock through which grains of magnetite and haematite are disseminated. The ore is crushed to a powder and electric magnets are used to separate the iron-bearing magnetite from the siliceous gangue. The beneficiation of taconite is a highly expensive process, justified by the great extent of the reserves of these low-grade ores, by the fact that railways, docks and other handling equipment are already in existence, at least in the Lower Great Lakes area, and by the approaching exhaustion of high-grade ores in the United States.[5]

Ores that have been subjected to such beneficiation processes, and many others that have not, are in too fine a state of division to be fed to the blast-furnace. They must be agglomerated into coarser particles that will not be blown out of the furnace with the top gases. This is usually done by sintering. The ore is heated—commonly by burning fine coke under a strong draught—and forms clinker-like masses, strong enough to withstand the pressures in a blast-furnace.

There are very few ores that could not be improved and transport charges reduced by beneficiation, though even in the United States no more than a third is subjected to this treatment.[6]

Beneficiation reduces the proportion of gangue, but it does not eliminate the very small quantities—often only traces—of sulphur, phosphorus and other such elements. Their presence is of the highest importance. They determine smelting methods, and those that pass into the metal influence strongly the types of refining or steel-making process that can be used.[7] The most objectionable of these is sulphur. A blast-furnace running at a very high temperature, and with a very basic slag, forces some to combine with the slag, but a trace usually passes over to the metal and may even survive the steel-making process. The faint smell of sulphur dioxide, which often pervades an iron works, is welcome to the furnace-man, because it shows that the sulphur is being oxidized and eliminated. The presence of sulphur in steel makes it 'hot-short', brittle when forged or heat-treated and liable to develop cracks. The furnace-man will often not accept ores containing over $0 \cdot 1$ per cent sulphur, and the large deposits of pyrites (iron sulphide, FeS_2), despite their quite high percentage of iron, are useless as ores. In the same way the furnace manager tries to avoid using coke which is high in sulphur, as the coke is usually the chief source of contamination for the metal.

Phosphorus is less harmful than sulphur, if only because it can be removed more easily. Most ores contain a trace of phosphorus, and this may amount to $1 \cdot 5$ per cent in such highly phosphoric ores as those of Lorraine. In addition, there is usually a trace of phosphorus in the coke. In the blast-furnace the phosphorus passes over to the metal. Its presence makes the iron very fluid and admirably suited for the manufacture of fine castings. But steel containing phosphorus is 'cold-short', that is, it breaks readily under pressure or stress. The puddling process had been able at least to bring the phosphorus content of wrought-iron within manageable proportions, but a high-phosphorus metal had defied the early efforts of the steel-makers (*see* page 24). The basic process remedied this, and, indeed, tipped the scales in the opposite direction. If iron has phosphorus at all, it had better have a lot if it is to be used in the Thomas process, in which the necessary heat is provided by the combustion of impurities in the metal. Even today Bessemer steel, made from non-phosphoric ore, is preferred for some purposes, and the world's scanty reserves of such ore have a disproportionately high importance.

Manganese, often present in very small quantities, is usually a highly desirable impurity. It gives strength to the metal and aids materially in the control of sulphur and oxygen. Indeed, manganese, in the form of ferro-manganese, or *spiegeleisen*, is added in many steel-making processes.

Titanium is sometimes injurious to the structure of the blast-furnace, but does not generally pass over to the metal in significant quantities. Vanadium, zinc, copper and arsenic may also occur in small quantities that are not usually injurious. Instances, however, are known where these occur in harmful amounts. It is usually desirable in such cases to mix the ore with ores of different composition, and so bring down the proportion in the blast-furnace charge.

Silica, alumina, lime and magnesia normally occur in much larger quantities and form a slag in the smelting process. Silica is the most obstinate of these owing to its high melting point. Lime has to be added and forms, along with the silica, a more readily fusible slag, which separates off from the iron. Some ores possess lime in sufficient quantities to make a fluid slag. But such 'self-fluxing' ores are few, and normally limestone has to be added to the furnace charge. Occasionally, as in Lorraine, it is possible to combine a lime-rich with a silica-rich ore to make a self-fluxing mixture. A considerable expense is saved if it is possible to use a self-fluxing ore or mixture of ores. The role of the acid alumina resembles that of silica, and it has to be neutralized and absorbed into a slag by the addition of the lime, but it is more easily fused than silica and, to that extent, is more easily managed. Magnesia, on the other hand, is basic in its reaction, and reinforces the action of the lime.

The slag plays a very important role in the smelting process, and a careful control of its chemical composition is essential if the quality of the iron is to be suitable for its intended uses. It is the medium through which those impurities in the ore and the fuel, which are not oxidized and removed with the exhaust gases, are withdrawn from the furnace. But the melting of the slag requires a great deal of heat; an abundant slag is wasteful of fuel, and for this reason the furnace manager keeps the slag ratio as low as possible. At the same time, he must add lime in quantities sufficient to neutralize the acid silica and to provide a basic medium that will absorb sulphur. The amount of silicon that may be allowed to pass over into the metal depends on the use to which it is to be put. Foundry iron can afford to be rich in silicon, and its smelting thus demands less lime. On the other hand, iron which is to be used in the converter should be low in silicon, as a result of using a more basic slag. In calculating the nature of the slag that will result from the use of certain ores, allowance must also be made for the fuel. All coke contains some ash, as well as sulphur. The ash, consisting mainly of alumina, is acid in its reaction, and thus requires lime for its elimination. The use of a coke with a high percentage of ash not only lowers the efficiency of the fuel, but, by necessitating a higher slag ratio, requires the expenditure of more heat to smelt each ton of pig-iron.

It is apparent that the grade of the ore is far from being the only factor of importance in assessing the value of an ore. Its physical texture and the nature and proportion of all the slag-making constituents are of vital importance. It is they which so often make it necessary for a country rich in ore to import other ores for blending purposes. Iron-smelting is an industry in which one does take coals to Newcastle.

World Iron-ore Reserves: An iron-ore is considered to be a deposit in which the metal is sufficiently concentrated and deleterious substances rare enough to make mining economic. It is impossible to calculate with any precision the size of the iron-ore reserves of the world. The recent opening up of the ore-field in northern Labrador reminds us that the age of major ore discoveries is not yet wholly passed, though the chances of making a find of revolutionary importance are diminishing rapidly. Further, the development and extension of beneficiation processes is changing our conception of what constitutes an ore. A generation ago, when the reserves of high-grade Lake Superior ore seemed abundant, the low-grade taconite would scarcely have come within the limits of the definition. Now, technological development, prompted by the hard necessity of getting more ore, makes it possible to mine, transport and smelt this low-grade material. The low-grade phosphoric ores of Lorraine were given their contemptuous epithet of *minette* by furnace-men who used only a richer, phosphorus-free ore which occurred in small quantities as a residual deposit over the limestone plateaus of eastern France. The latter is exhausted, and *minette* is an ore *par excellence*.

Lastly, the small size of very many deposits removes them from the sphere of practical mining for reasons outlined above. It is not surprising, then, that estimates of ore reserves differ widely, nor that there is a tendency for the totals to increase as ores of lower grade come to be included.

In the following paragraphs the nature and extent of the ore deposits in the more important iron-producing countries are reviewed.

United Kingdom:[8] The British iron-ores belong to three types. Most valuable and least abundant are the phosphorus-free haematites of Cumberland and Wales. It was on these that Bessemer established his converter process, but today they make up only about $2\frac{1}{2}$ per cent of the ore mined. The coal-measures ores, the clayband and blackband, were, at least locally, of considerable importance, but output has been declining for many years, and their contribution is even smaller than that of the haematite ores. On the other hand, the bedded Jurassic ores of the East Midlands, Lincolnshire and Yorkshire have grown in importance, and in 1951 accounted for 97 per cent of the ore by weight.

Estimates of the volume of these ores differ widely. A figure of 2,360 million tons, all of it Jurassic except some 6 million of haematite, may be unduly optimistic, but there is a certain reserve for 150 years at the present rate of consumption. These Jurassic ores can generally be worked in open pits (*see* page 87), but their grade is low, and averages only 27·3 per cent of iron. They have a high phosphorus content, are rarely self-fluxing and commonly require to be sintered before being put into the furnace.

France and Luxembourg: The *minette*, similar in geological age and chemical composition to the English Jurassic ores, is the most abundant ore deposit in Europe, and one of the largest in the world. Very little of it is now obtained from open cuts, and most is obtained from underground workings, where it is blasted and machine-loaded on to trucks. It has certain advantages over the English Jurassic ore.

The actual mining costs may be a little higher, but as it is obtained from underground workings, the problem of the destruction and reconstitution of the land surface does not arise. Furthermore, siliceous ores occur in close proximity to calcareous, so that by blending the two a self-fluxing ore can be put into the furnace.

In addition to the bedded ores of Lorraine, France has also reserves of similar ores in many parts of the Paris Basin and in Normandy. Though no longer mined, they were formerly of considerable importance, and have been eclipsed only by the greater abundance and ease of working of the *minette*.

Like England, France has also deposits of high-grade haematite, low in phosphorus. But the deposits, mainly in the French Alps, the Pyrenees and in the Breton peninsula, are individually small, and for that reason are little worked today. They were, however, formerly of great importance on account of the high quality of the metal smelted from them.

Germany: The iron-ore resources of Germany belong to two contrasted types. Of the greatest importance in the past have been those ores included in the massifs of Paleozoic rock which form a belt from west to east across Germany and continue into Bohemia. Many of these were of high grade, relatively free of phosphorus and rich in manganese. The Siegerland, deriving much of its ore from the Erzberg at Müsen, was one of the most important centres of iron-smelting in Germany, until it was eclipsed by the Ruhr. The Harz Mountains and the Thuringian Forest also had numerous deposits, much used in earlier times.

Today, however, German iron-ore mining is largely concentrated in the plains of North and South Germany. There are large deposits, though less abundant than those of France and England, of bedded Jurassic ores. Their extraction and use presents the same kind of problems as the mining and smelting of *minette* and scarplands ore. But such ore from Lower Saxony now makes a very significant contribution to Germany's total ore consumption.

Sweden: Iron-ore production is today dominated by the vast reserves of Swedish Lapland, but until the railways were completed, late in the nineteenth century, linking these with the sea, the chief sources of ore were in Central Sweden.[9] Most of the Swedish deposits are either of igneous origin, or closely associated, as residual or derived ore masses, with igneous activity. A consequence of this is the predominance of the richer magnetite among the deposits. The ores of Central Sweden are relatively free of phosphorus and sulphur, a reason for the high reputation enjoyed by Swedish pig-iron for many centuries, but the rich magnetites of Lapland can be used only for smelting basic pig-iron. The ores from the Bergslagen, as the mining district of Central Sweden is traditionally called, have been used since the thirteenth century at least. But the reserves here are not large, output is today quite small and is reserved for the needs of the Swedish industry.

The Lapland deposits were known at least as early as the seventeenth century, but lack of transport and the inhospitable region in which they lay prevented their use until the later years of the nineteenth. They occur as a number of large masses or concentrations of magnetite in the granitic rocks of northern Sweden. They are somewhat harder than the surrounding rocks, and generally form low rounded hills. This greatly facilitates mining and, though deep mining is practised, much of the ore is obtained from open cuts. The chief ore masses, Kiruna, Gallivare-Malmberget and Koskullskulle, are linked by railway both with the Baltic coast and with Narvik, on the Atlantic coast of Norway. Ore shipments by the Baltic ports are interrupted during the winter, but take place from Narvik throughout the year.

There are other ore-bodies in southern Lapland, with Boliden as the most important at present. They lie closer to the Baltic coast than those of northern Lapland, and ore is shipped from neighbouring ports during the open season.

The Swedish deposits are of great significance not only on account of the great volume of the deposits themselves, but because their grade is high enough to justify expensive transport to European and even American markets, without preliminary concentration. A large part of the West German smelting industry and a part of that in the United Kingdom and France has been tailored to fit the needs of Swedish ore.

Norway: The Norwegian deposits resemble those of Sweden, but are much less extensive and of a lower grade. The ore is mined at several points close to the coast in northern Norway, and is shipped to furnaces in Western Europe. The most extensive deposits, and also the most productive mine, lies inland from Kirkenes, very close to the boundary between Norway and the Soviet Union.

Eastern Europe: In Eastern Europe, where an industrial revolution of great importance and magnitude is at present taking place, the deposits of worthwhile ore are small and few. Poland and Czechoslovakia, the most richly endowed of these countries, are obliged today to import a large proportion of the ore consumed, and as their smelting industry expands they will become increasingly dependent on foreign sources of ore. In Poland the most abundant ores are the *minette*-type deposits of the Krakowskie Jura. In Czechoslovakia deposits are worked in the Hercynian rocks of Bohemia and in the Carpathian mountains of Slovakia. Deposits in Hungary, Romania and Bulgaria are small and scattered and of very small economic value. Only Yugoslavia has more abundant deposits. The ores of Croatia—mainly limonite—and of Bosnia and Serbia are adequate to support a small iron-smelting industry.[10]

Southern Europe: The ores of Spain and Italy have enjoyed a high reputation, but reserves are at present small, and mining here cannot look forward to a very bright future. The haematite ores of the Bilbao district of northern Spain are very low in phosphorus, and were formerly in great demand for smelting Bessemer iron, but only very small quantities are now exported. Ores are found in southern and eastern Spain, and though they have been mined and exported, they are not of great importance today.

In Italy the most renowned ores were those of the island of Elba, which were used at least as early as the period of the Roman Republic. They continue to be mined, but the deposits are today near exhaustion. Italian ores, like those of Spain, occur in relatively small deposits, many of which cannot for this reason be worked economically.

Middle East: Large workable deposits of iron-ore appear to be almost completely lacking. Only Turkey has reserves which are at present mined. The Divrik ore-field of eastern Anatolia is not large, but it is the only domestic source used in the supply of ore to Turkey's new iron-smelting industry.[11]

Soviet Union: It had been supposed that the Soviet Union was not well endowed with iron ore,[12] but recent evidence is that it has more than its proportionate share of the world's resources. The ores are heavily concentrated in the Ural mountains and European Russia, and 'in the whole expanse of the Soviet Far East, no deposit of high-grade iron-ore has yet been found'.[13] Even in European Russia the large and widely publicized deposits of high-grade ore at Magnitogorsk and Krivoi Rog are nearing exhaustion and the Soviet Union is obliged to rely more and more on lower-grade deposits.

The Krivoi Rog ores of the southern Ukraine are still the most important in the Soviet Union. They occur in a narrow band, some thirty-five miles in length and five in breadth, enclosed within the great bend of the Dnieper river. The ores are mainly haematite of fairly high grade. The ore is phosphoric and has a high silica content, thus necessitating a large addition of lime. It is a very friable deposit, easy to work, but requires to be sintered before smelting.

The Krivoi Rog deposit was opened up in 1881, and since that date has supported by far the largest part of the Russian smelting industry. When the First World War began, about 70 per cent of ore mined in Russia came from Krivoi Rog. This proportion has since declined but the mines probably still contribute nearly a half. There has, however, been a marked deterioration in the quality of Krivoi Rog ore, both in grade and in texture, as the best deposits are exhausted and replaced by low-grade ferruginous quartzites. Second in importance to Krivoi Rog is the 'iron mountain' of Magnitogorsk. The ore-body is a mass of high-grade magnetite. In the early 1930's it began to be heavily exploited to supply both the local furnaces and those erected in Kuznetsk, in western Siberia. There can be no doubt that here, as at Krivoi Rog, the best of the ore has already been worked. For the future, ore of a diminishing quality will be mined at an increasing cost. At present Magnitogorsk produces about 15 per cent of the total Soviet output and by 1980 the deposit is likely to be completely exhausted.

This diminishing importance of the two primary ore-bodies of the Soviet Union is likely to continue, and the Russians are giving increasing attention to other ore deposits, whose use would not necessitate very long rail hauls. Among these are the Kertch deposits of the eastern Crimea. They are very abundant, but of fairly low grade, phosphoric and rich in silica, and, like the Krivoi Rog ores, are very powdery and difficult to handle. Problems of preparing and blending the ore for smelting are serious, but, nevertheless, two smelting works have been established to use it.

In the centre of European Russia are numerous deposits mainly of brown ore, or limonite. These have been worked at intervals for over two centuries, but they are in general poor ores, and the costs of mining and smelting them are high. Among these scattered ore-bodies is a magnetite deposit of exceptional size. It lies near Kursk, is of a fairly low grade and is hard and, for technical reasons, difficult to mine. But magnetite is more easily concentrated by magnetic means than other ores of iron, and, as M. Gardner Clark has observed, 'If this quartzite ore can ever be made suitable for the smelting of pig-iron by beneficiation, then Soviet blast-furnaces will have at their disposal unlimited ore supplies, located in the industrial heart of the Soviet Union'.[14] Present plans, however, do not seem to include the exploitation of Kursk magnetite.

In the extreme north-west of the Soviet Union are deposits of low-grade magnetite, resembling those worked in northern Norway. These are now being used, though perhaps only on an experimental basis, as they require concentration and their phosphorus content is almost unmanageably large.

In the Ural mountains are numerous ore deposits, in addition to that at Magnitogorsk. Some of them are small, and many contain ore that can be smelted only with great difficulty. Several are exploited today, and more of them were used in the days of the eighteenth- and nineteenth-century charcoal iron industry. With the exhaustion of at least the better ores of Magnitogorsk, the Russians are again turning their attention to these lesser deposits. The only really large find is at Kustanay in Kazakhstan.

Siberia is poor in rich iron-ore deposits. The most valuable are those to the south-east of Novosibirsk. They are not abundant, and contain inconveniently large amounts of sulphur, silica and even zinc. They are, by themselves, inadequate to support the large smelting industry that has been established in their midst. There are small deposits in Transcaucasia, in Soviet Central Asia, and in the Soviet Far East. They are used for lack of better, but by themselves could not possibly justify the setting up of smelting industries. Soviet policy has been to disperse its iron and steel works; the distribution of ore resources suggests very strongly that the industry should be concentrated in two or three areas. (*See* Chapter 7.)

India and Pakistan: The only significant deposits in Pakistan lie in the inaccessible mountains of the north of the State. Needless to say, they are not exploited. India, by contrast, has a wealth of good and easily accessible ore.[15] The best of this lies in the districts of Singhbhum and Mayurbhanj, on the borders of Bihar and Orissa, to the west of Calcutta. It is a sedimentary ironstone of pre-Cambrian age, and resembles in its composition and occurrence the deposits of Lake Superior. The iron content has in many parts been enriched by the leaching of the silica, and the result is an ore-body of more than usual richness. The ores yield a phosphoric iron, but their sulphur content is low. The chief deposits occur in a series of parallel ridges, of which the largest is over thirty miles in length and at most a mile in width. The deposit as a whole is undoubtedly the richest in Asia and must rank as one of the greatest in the world.

Other deposits, similar in origin and composition to those of Bihar and Orissa, though very much smaller in total reserves, occur in Bombay, Hyderabad and in the hills along the northern margin of the Deccan. These have been used, though large-scale mining is still confined to the Singhbhum and Mayurbhanj deposits.

South-East Asia: There are many deposits of iron-ore in South-East Asia; some of them lateritic accumulations on the surface of the ground, others associated with granitic intrusions or regions of ancient rocks. Unfortunately most are small, and it has been said that none of those in Burma could support a modern smelting industry.[16] Ore-bodies in Malaya and Viet-Nam are larger, and some of them are now being worked and the ore exported. Indonesia also has scattered deposits, many of them lateritic and of low quality. But in the whole of South-East Asia, only the Philippines have ores of the amount and quality required by a modern iron industry. These ores include high-grade magnetite of igneous origin, and vast quantities of poorer, lateritic ores. Ore is mined and exported, but has not yet been used as the basis of a local smelting industry on the modern scale.

China: Despite its great area, China is not really rich in iron-ores, and its total known reserves are a great deal smaller than those of France or England. Nevertheless, further discoveries of ore-bodies may be expected in a country as little explored geologically as China. The most valuable deposits are in Manchuria.[17] These resemble the Lake Superior deposits in origin and occurrence, but are very much smaller in extent and lower in their iron content. Though high-grade haematite and magnetite ores occur, many of the deposits resemble the Minnesota taconite, and require concentration before they can be used in a blast-furnace.[18]

It is estimated that only about a third of the Chinese ore reserves are to be found in the rest of the country.[19] These are widely scattered in a great number of relatively small deposits. Many are found in rugged country, almost completely lacking in modern means of transport. In scarcely any of those that have been prospected and surveyed is the volume of ore sufficient to supply a modern blast-furnace works for its normal working life. Many are too small to justify the installation of modern mining, ore-handling and concentrating plant. But this is China; there is no reason why many of these ores should not be worked manually and smelted in small furnaces, inexpensive to construct but extravagant of labour. China could not, on this basis, compete with the West; but there is no reason why she should not satisfy a much larger home demand than exists at present.

Numerous deposits occur in Korea, especially in northern Korea, and the Musan deposit, close to the Manchurian border, is large. Several ore mines were opened by the Japanese close to the west coast, from which shipments could be made to Japan. Korea can support a small smelting industry, but long-term plans can be laid only on the basis of the Musan ores.

Japan: The islands of Japan appear to have a large number of small ore deposits, perhaps because the country has been more carefully prospected than other parts of Asia. But the majority have each less than a million tons of ore, and much of this is of medium or low grade. Thus, domestic supplies can at best only supplement the import of ore, and, if used exclusively, could not supply the present Japanese smelting industry for more than a year or two.

Australia and New Zealand: The continent of Australia is comparatively rich in iron-ores, and most of them, unlike those of East Asia, occur in large compact masses, suited to mining by the most modern methods. The greatest commercial importance at present attaches to the deposits of the Middleback Range of South Australia. These resemble the Lake Superior deposits. They occur as folded sedimentaries of pre-Cambrian age in a series of rounded hills. These hills lie strung out in a chain about forty miles from north to south, and twenty-five miles inland from Whyalla, on Spencer Gulf. The most northerly of these iron ranges is Iron Knob; the most southerly, Iron Duke, and between lie Iron Monarch, Iron Baron and other similarly designated hills. Reserves are large, the iron content of the ores is generally high, and their location convenient for sea transport.

Even larger in extent, though so far but little used, are the deposits of Western Australia. These also occur as beds in the pre-Cambrian massif of Australia. In general they contain over 60 per cent iron, though there are also large quantities of lower-grade ore. These deposits occur along the northern coast of Western Australia, and at Yampi Sound they are now being mined and shipped to smelters in New South Wales. In the interior of Western Australia, often remote from means of transport, are many other deposits of rich ore, which make this one of the better-endowed regions of the world.

Deposits in Queensland and New South Wales are much smaller than those in South and Western Australia, and very little use has hitherto been made of them. But reserves are considerable, and, if they had occurred instead in Japan, would long since have been fully exploited.

New Zealand possesses considerable reserves of iron-ore, chiefly iron sands, which contain also titanium. No attempt has yet been made to work them commercially. The island of New Caledonia has extensive deposits of lateritic ore. Though the deposits are worked chiefly for their nickel content, a small export of iron-ore takes place to Australia.

Africa: Geological surveying has made too little progress in Africa for a reliable estimate to be made of its resources in iron-ore. The fact that two-thirds of the known resources are in the Atlas region of North Africa and in the Union of South Africa and Southern Rhodesia is a measure rather of our ignorance of the rest of the continent than of its inherent poverty.[20]

Earliest to be developed on a modern scale were the bedded ores of Algeria and of other parts of French North Africa. They were good haematites, much in demand in Europe in the second half of the nineteenth century for making Bessemer iron, but reserves of these ores are now not amongst the larger deposits of the continent.

Africa is a tableland of ancient rocks, among which pre-Cambrian sedimentaries are prominent. Ore-bodies, broadly similar to those of Australia or Lake Superior, are numerous and some are very large. Deposits are particularly rich in the Union of South Africa, and some of the oolitic deposits, resembling *minette*, of the Transvaal, are very large indeed.

Ore is mined and smelted locally in both the Union of South Africa and in Southern Rhodesia, but elsewhere in the continent ores are mined for export only. Naturally only those deposits favourably situated in relation to ocean transport are worked, and hitherto interest has concentrated on the fairly accessible ores of West Africa. Here there are both extensive surface deposits of laterite and also beds of pre-Cambrian ores. In recent years lateritic ore has been exported from Conakry, in Guinea. A richer and purer ore is obtained from the Bomi hills in Liberia, where the Republic Steel Corporation of the United States has acquired extensive holdings, and is shipping ore to North America.[21] Sierra Leone has extensive deposits inland from Freetown, which are now being mined for export. But one of the largest and richest deposits is near Fort Gourand in French Mauretania. These deposits were discovered only after the end of the Second World War, and their exploitation is being held up by lack of developed transport.

North America: The American continents may well prove to contain by far the largest iron-ore reserves in the world today. For many decades now the United States has been the world's biggest producer of ore, and, though some of the better deposits are nearing exhaustion, there is no sign of any change in America's relative position. The iron-ore resources of the United States are dominated by the vast reserves of the Lake Superior region, of the southern Appalachians and of the Adirondacks.

The Lake Superior deposits have been exploited now for over a century, and altogether some 3,000 million tons of ore have been shipped in this period to the smelting works in the north-eastern United States.[22] The ore deposits consist of gently folded beds of haematite, magnetite and other iron compounds, contained in a massif of pre-Cambrian rocks. The iron-bearing deposits form low hills—the iron ranges—at the surface, where the ore can be worked by open-pit methods. But in all cases the ore dips beneath the surface and has to be reached by shafts. Published estimates of existing reserves suggest that three-quarters of the whole deposit has already been mined. Such estimates, however, are published for taxation purposes, and consequently do nothing to exaggerate the size of the reserves. However conservative these estimates may be, it seems reasonably certain that a great deal less than half of the high-grade ore remains.

The ranges lying to the south of Lake Superior contain relatively small reserves, and most of the remaining ore is in the Mesabi Range, to the north-west of Duluth. Closely associated with the Mesabi haematite are very large deposits of a low-grade—about 27 per cent iron—ore which has come to be called taconite (see page 34). Where the iron contained in it is in the form of magnetite, the ore has only to be crushed and magnetically separated. But haematite has first to be roasted to alter the Fe_2O_3 into the magnetic form of Fe_3O_4. If the taconite is considered, there remains far more iron in the Lake Superior region than has hitherto been removed from it.

The iron-ore deposits of the Adirondack mountains of New York State resemble in some ways those of northern Sweden. They occur as pre-Cambrian rocks, and they contain a high proportion of magnetite, which lends itself to easy concentration. On the other hand, the ore-bodies are not large and some are nearing exhaustion.[23]

The third major source of ore in the United States is in the southern Appalachian mountains. Bedded oolitic ores, similar to the *minette* of Lorraine, extend over a considerable area of northern Alabama and adjoining parts of Georgia and Tennessee. Total reserves are very large and probably exceed those of good grade Lake Superior ores. The Clinton ore formation is mined near Birmingham (see page 133) and supplies the local smelters. It is a fairly low-grade ore—from 31 to 39 per cent—and cannot profitably be transported far without concentration. It is also a phosphoric ore, but the phosphorus content is not usually high enough for the converter process to be used successfully.

There are smaller ore-bodies scattered through the Appalachian mountains. They have been used since the early eighteenth century to supply small blast-furnaces, but these deposits have today only a slight importance, and have been abandoned except at a few sites in Pennsylvania and New Jersey.

Iron-ore deposits west of the river Mississippi have received less attention than those to the east, chiefly because the need for them has been less. There is a deposit of magnetite and haematite, of the Kiruna type, in Iron Mountain, Missouri, but, apart from this, there are few significant deposits in the Plains States. In the Mountain States reserves are large but little used. Utah has large masses of good haematite and magnetite, and, should the need arise, large quantities could be obtained from Wyoming, Montana, Nevada and California. Present production from these western sources does not exceed 10 million tons annually, but plans are said to be in hand for a considerable expansion of output.[24]

Canada: With her very large area of pre-Cambrian rocks, Canada has large reserves of ores of the Lake Superior type. Such deposits have been prospected and are now being worked to the north of Lake Superior, especially at Steep Rock and Michipicoten. The largest of such deposits yet located lies on the borders of Quebec and Labrador. It is too early to assess the size of this reserve of ore, which may prove to be the equal of Mesabi itself, but already large quantities of ore are being shipped from Seven Islands, on the Gulf of St. Lawrence.

Most of the known Canadian deposits are of pre-Cambrian geological age, but the very large reserves at Wabana, on Bell Isle, off the east coast of Newfoundland, are much later in date and consist of oolitic ore similar to *minette*. They have a higher iron content than most oolitic ores, and their location, on a small island in Conception Bay, makes it relatively easy to export them.

There are numerous ore bodies in the Canadian Rockies associated generally with igneous rocks that were intruded when the mountains were formed. It has not yet been demonstrated that these deposits are important additions to Canada's mineral wealth. Similar deposits have been found in Alaska, but their nature and extent are not yet fully known.

Mexico and Central America:[25] Many ore-bodies have been

found in Mexico, but the lack of careful geological study makes their evaluation impossible. It would appear, however, that several of them are large and of a high quality. In the light of present knowledge the most abundant reserve is Cerro de Mercado, in the State of Durango. It is a segregation deposit in extrusive rocks of Tertiary age. This ore-body is used in the Mexican blast-furnace works, but many other good-quality deposits are too remotely situated to be exploited in the near future.

The island of Cuba has extensive deposits of lateritic ore, formed on basic igneous rocks. They can be beneficiated with comparative ease, but, like many lateritic ores, contain enough nickel and chromium to pose difficulties in smelting and steel-making. In addition, Cuba has small deposits of non-hydrated ores which have also been mined and exported. Puerto Rico also has large deposits of lateritic ore, as well as small amounts of magnetite, but iron-ore deposits in other parts of Central America and the West Indies do not appear to be significant.

South America: Any consideration of the iron-ore reserves of South America is necessarily based on quite inadequate data. Geological surveying has made so little progress in most parts of the continent that it seems certain that reserves remain to be discovered. Furthermore, the development of the iron-smelting industry has so far been on too small a scale to stimulate further surveying. At present the bulk of the ore appears to lie in Brazil, Venezuela and Chile.

Brazil not only dominates South American production but may be found to have the largest deposit of high-grade ore in the world. The massive deposits in the province of Minas Gerais are a high-grade haematite of sedimentary origin. In the best of them the silica has been leached, leaving a soft and easily worked ore with up to 68 per cent of iron. The actual reserves of this ore have been put at 16,250 million tons, and the potential reserves at more than twice as much more. This does not exhaust the ores of Brazil. The Mato Grosso deposits, on the far western border of the country, are very large though not yet exploited. And there are other, smaller deposits in São Paulo, Bahía and the lower Amazon valley.

Venezuela has become important as a source of iron-ore only in recent years. In the mountains of the interior of the country are large deposits of sedimentary ores, enriched, like those of Minas Gerais, by leaching. They are worked at Cerro Bolivar and El Pao by companies controlled by two of the large United States steel corporations. The ore ranges up to 68 per cent in metal content, and the deposits are soft and easily worked in open cuts. But the workings lie up to 300 miles from the coast, and the complexities of land and water transport greatly heighten the cost of the ore. Nevertheless, all the ore from these sources is exported, and most of it goes to the United States.

The only other country in South America with important reserves of iron-ore is Chile. These are not as large as those of Brazil and Venezuela, but appear to be mostly high grade and relatively free from harmful impurities. They lie, furthermore, close to the coast, and their export does not pose any serious problems. In 1912 the El Tofo concession, in the southern part of the Atacama region of Chile, was acquired by the Bethlehem Steel Company, and since 1922 ore from this source has been used to supply the company's works at Sparrow's Point, near Baltimore. This is probably the longest route regularly used from iron-ore mine to smelter, and is justified only by the high quality of the ore and the favourable mining conditions in Chile.

Summary: Many attempts have been made to assess the world's resources in iron-ore. They vary within wide limits. Not only has the essential geological survey work not been done in many regions of the world, but the basis of assessment varies from one country to another. What in one is counted only as potential reserve might in another be quoted as actual, and there is no measure of agreement in deciding what is the lowest metal content that may still constitute an ore. Lastly, just as the advancing scale and complexity of mining have already rendered many small deposits economically useless, so, in the same way, ores which at present may be too low in iron or too high in impurities to be workable may be brought within the range of profitable mining.

The distribution of iron-ore deposits cannot be shown precisely on so small a map as Fig. 1. It is likely to be modified as new discoveries are made and further light is thrown upon those already known. The basic geographical pattern is, however, unlikely to be greatly changed, and the ironic picture of relative poverty among the great industrial nations and relative wealth among the under-developed is likely to persist.[26]

The following table, based on the same sources, gives a brief and crude picture of the same situation:

	Proved Ores	Potential Ores
	(in millions of tons)	
Great Britain .．	3,760	5,800
Western Europe:		
France	6,560	5,120
Germany	1,510	1,630
Sweden	2,400	1,170
Other countries in non-Com-		
munist Europe	1,830	2,650
Communist Eastern Europe ..	430	570
U.S.S.R.	3,140	8,310
India and Pakistan	5,285	17,691
China	2,160	—
Rest of Asia	2,210	421
Africa:		
North Africa	95	150
Union of South Africa ..	280	3,600
Rest of Africa	1,020	2,460
Australia	984	67
North America:		
U.S.A.	5,200	105,200
Canada	2,944	6,704
Central America and West Indies	213	3,053
South America:		
Brazil	17,505	50,862
Venezuela	500	2,220
Chile	77	616·8
Rest of South America ..	80	1,852
TOTAL	58,183	220,146·8

These totals, which may well prove to be an underestimate, seem so huge that man might afford to be prodigal of his resources. But much of these reserves may never, with the technology that he has at present at his command, come within the field of profitable mining. Furthermore, it must be remembered that he has already taken some 3,000 million tons of ore from the deposits grouped around Lake

Superior alone, and that, to produce the pig-iron which he has made since 1870, he has taken over 10,000 million tons of ore from the earth. He cannot afford to be too extravagant even with this seeming abundance of iron-ore.

Iron-ore is only one element in the blast-furnace charge. No less important is the fuel without which the ore could not be reduced to a metal and the limestone which allows the impurities to be removed as a liquid slag. To these, and also to that vital question of the re-use of old metal, we now turn.

[1] There is an excellent review of the subject by Frederick G. Percival 'Nature and Occurrence of Iron Ore Deposits', in *World Survey of Iron Ore Resources*, United Nations, 1955, 45–76. This has a useful bibliography. See also *The Making, Shaping and Treating of Steel*, seventh edition, United States Steel, Pittsburgh, 1957, Chapter 6, 'Iron Ores'.

[2] *Description des Arts et des Métiers*, Paris, Volume XIX, 1762.

[3] Based on *Survey of World Iron Ore Resources*, Part II, 'Appraisal of Iron Ore Resources'.

[4] John D. Sullivan, 'Beneficiating Iron Ore', *Survey of World Iron Ore Resources*, 106–21; *Low Grade Ores: A Survey of American Research Methods*, European Productivity Agency, OEEC, 1958.

[5] *Mineral Facts and Problems*, United States Bureau of Mines, Bulletin 556; Clyde F. Kohn and Raymond E. Specht, 'The Mining of Taconite, Lake Superior Iron Mining District, *Geographical Review*, XLVIII, 1958, 528–539.

[6] *Materials Survey: Iron Ore*, United States Bureau of Mines, 1956.

[7] Martin Wiberg, 'Relation of Type of Ore to Smelting Processes', *Survey of World Iron Ore Resources*, 122–45.

[8] 'Britain's Iron Ore Resources', *Monthly Statistical Bulletin*, B. I. & S. F., XXVII, October, 1952.

[9] A. F. Rickman, *Swedish Iron Ore*, Faber & Faber, 1939; Gunnar Lowegren, *Swedish Iron and Steel*, Stockholm, 1948.

[10] I. Avsenek, *Yugoslav Metallurgical Industry*, New York, 1955.

[11] *Survey of World Iron Ore Resources*, United Nations, 1955.

[12] Demitri B. Shimkin, *Minerals: A Key to Soviet Power*, Harvard University Press, 1953, 33–51.

[13] M. Gardner Clark, *The Economics of Soviet Steel*, Harvard University Press, 1956, 187.

[14] *Op. cit.*, 160.

[15] See *Survey of World Iron Ore Resources*, 270–96, for a detailed survey of Indian resources; also Pradyumna P. Karan, 'Iron Mining Industry in Singhbhum-Mayurbhanj Region of India', *Economic Geography*, XXXIII, 1957, 349–61; and John E. Bush, 'The Iron and

Steel Industry in India', *Geographical Review*, XLII, 1952, 37–55; J. Coggin Brown and A. K. Dey, *India's Mineral Wealth*, Oxford, 1955.

[16] *Survey of World Iron Ore Resources*, 297.

[17] M. ErSelcuk, 'The Iron and Steel Industry of China', *Economic Geography*, XXXII, 1956, 347–71; J. S. Lee, *The Geology of China*, London, 1939.

[18] *See Survey of World Iron Ore Resources*, 311–326, and H. Foster Bain, *Ores and Industry in the Far East*, New York, 1933, 83–106.

[19] *Ibid*, 311.

[20] Based on *World Iron Ore Resources and Their Utilization*, United Nations, 1950, II; A. Williams Postel, *The Mineral Resources of Africa*, African Handbooks: 2, Philadelphia, 1943; *The Mineral Resources of the Union of South Africa*, Union of South Africa Department of Mines, 1936, 201–210.

[21] *Materials Survey: Iron Ore*, United States Department of the Interior, 1956, V, 93–95.

[22] A technical account of these ores, by Carl E. Dutton, is given in *Survey of World Iron Ore Resources*, 188–92. There is a bibliography on pages 206–8.

[23] *See* Carl E. Dutton, *op. cit.*, and Sven A. Anderson and Augustus Jones, 'Iron in the Adirondacks', *Economic Geography*, XXI, 1945, 276–85.

[24] *Iron*, Bulletin 556, Bureau of Mines, 1955, 24.

[25] *A Study of the Iron and Steel Industry of Latin America*, Volume I, United Nations, 1954.

[26] H. M. Mikami, 'World Iron Ore Map', *Economic Geology*, XXXIX, 1944, 1–24.

3

THE MODERN IRON AND STEEL WORKS

THE iron and steel works is the means by which the ore is converted into metallic iron with predetermined physical and chemical properties. The process is chemically a relatively simple one, and varies in its execution to only a very slight degree over the world. The plant on the other hand is complex and represents so large a capital investment that the scale of the operation is of immense importance. Both raw materials and finished products are bulky and awkward to transport, so that the location of works poses a difficult and quasi-mathematical problem. Lastly, the plant covers a large area and usually needs room to expand, and it must be well placed for handling the bulky materials brought to it by rail or water. It must be able to attract labour and to supply the water needed in very large quantities for cooling purposes.

The uses for cast-iron have diminished in recent times, and the economies that result from passing the pig-iron direct to the steel furnace are such that the integrated iron and steel works is now normal. There remain, however, a few blast-furnace works which market only pig-iron or iron castings. Some of those of the Siegerland are examples. But these works are generally small and sometimes obsolete. At the opposite end of the process, steel works exist without blast-furnace very much more frequently. There are several reasons for this, and foremost amongst them is the fact that most steel is made wholly or in part from scrap. It is usual for the electric furnace to use only scrap, and the open-hearth charge is commonly half scrap. The heavy dependence of electric steel production on electric power—frequently hydro-electric—means that this branch of the industry moves towards the source of power and is almost wholly divorced from the smelting branch. The relatively low freight charges on pig-iron, as contrasted with very much higher rates on rolled steel goods, sometimes make it desirable to establish steel-making and rolling works near large consuming centres, where also the supply of scrap is likely to be greatest.

The Integrated Works:[1] But the integrated iron and steel works is the normal means of production, and almost all the developments in the industry in recent years and those planned for the future call for this type of equipment. In short, an integrated works consists of blast-furnaces, steel works and rolling-mills. To these are frequently added coke-ovens and such plant as may be needed to prepare the ore for the furnace and to process the by-products of the coking and metallurgical processes. There may also be a steel foundry and a press or forge works in which steel ingots are pressed or hammered to the required shapes. In the organization of steel production, how-ever, it is usual to find the foundry, press and forge works which are, from the nature of their products, less susceptible of large-scale organization more closely associated with the market than with the smelting and steel-making end of the process.

Most steel products are, however, rolled from steel ingots to standardized sizes and shapes. At a large works, the steel is poured into ingot moulds. After the ingot has solidified and been stripped of its mould, it is reduced to more manageable proportions in a blooming- or slabbing-mill. The blooms are, after reheating in a soaking-pit, rolled to bars, billets, rails and other 'shapes', while the slabs are rolled to sheets and strips.

In the United States, and in the more progressive works else-where, the rolling often is continuous. In other words, the slab or bloom moves continuously in the *same* direction through a series of rolls that is long enough to reduce the metal to its final thickness and size. Previously the movement of the metal had to be reversed, so that it passed back through the same rolls again and again. There were devices for speeding this and for obviating the need for the engines which powered the rolls to stop and reverse at each change. But the older process was slow and expensive, especially of labour. The continuous process, on the other hand, calls for engineering skills of a high order. The ratio of the roller speeds to one another must be precisely determined and accurately maintained. Further-more, one cannot have a *small* continuous mill.[2] It must be a large piece of equipment, with a large output, and all other parts of the plant must be geared to it.

There is a thickness of metal below which hot rolling is not practicable. In the production of very thin sheet, such, for example, as is used in making tinned containers, the final rolling of the metal is at a low temperature. This not only gives a smoother surface, which is desirable for tinning, but permits of a much closer control over thickness. The 'cold-drawing' of wire is also practised to bring it down to very small thicknesses. The cold processing of steel is also frequently divorced geographically from the smelting and steel-making processes.

The diagram on page 58 shows somewhat schematically the processes in a modern integrated works. The materials 'flow' from one end of the works to the other, and it is clearly desirable that consecutive processes should be carried on next to one another in order to minimize handling and movement. Ideally the plant needs a large area of level ground over which it can stretch from the fuel and ore stockpile at the one end to the storage space for rolled goods at the other, and the two ends must be closely linked with transport facilities. In addition, both the smelting and steel-making processes generate large quantities of slag. Some is used in by-product plants —cement, paving and building blocks, phosphatic fertilizer—but most has no commercial value. Space is needed to tip this useless material. Most long-established iron-working areas can show their ugly grey-brown slag piles, over which little can be made to grow. The works in northern Indiana are exceptional in their freedom to tip such material into Lake Michigan, and thus to create new land for themselves.

With very few exceptions, modern developments in iron and steel manufacture are on virgin sites, which allow for almost unlimited growth: the Abbey works at Margam, Nowa Huta, the Fairless works near Philadelphia and Volta Redonda in Brazil. By contrast, old and congested sites have been abandoned, such, for example, as those in the built-up areas of Chicago and some in Pittsburgh, in Yorkshire and in the more highly urbanized parts of the Ruhr. Many others, though generously planned a generation ago, now find themselves hopelessly cramped. Even some of the lakeside works of the Chicago area have been enveloped by the city's recent sprawl and

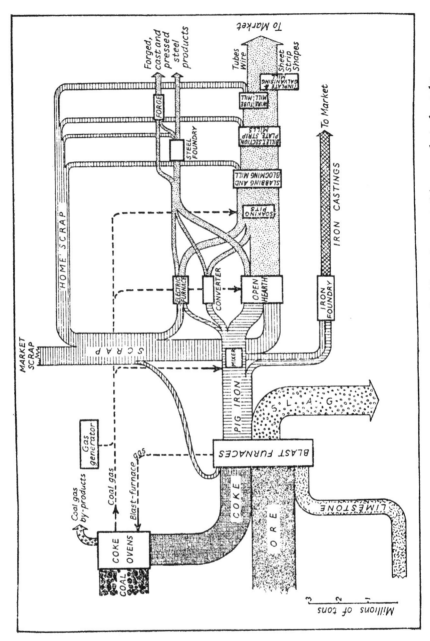

Fig. 3. Flow diagram of materials used in a typical large integrated iron and steel works. The scale is approximate only

find themselves in considerable difficulties even in that land of the wide open spaces.

The Problem of Location: The raw materials from which pig-iron is smelted and steel is made are bulky and heavy, and the cost of their transport to the place of manufacture is a major item in the total cost of production. Indeed, it has been estimated that a third of the cost of pig-iron and two-fifths of the price of steel-mill products represent the transport charges on materials.[3]

Yet the ore, fuel, limestone and scrap which flow to an iron and steel works generally travel at a relatively low freight-rate per ton-mile. If they are carried by water in 'bulk-carriers' the cost of transport is even lower. The cost of handling rolled, cast or forged goods is normally a great deal higher. Not only do the railways charge a higher freight-rate but the economies to be derived from bulk handling and bulk carriers are usually lacking. The ratio between the costs of assembling the raw materials and of despatching mill products to the market clearly varies with distance, mode of transport and type of product, and also with the type of contract that has been made with the carriers. For most industrial sites, however, the cost of assembling raw materials exceeds by a wide margin the cost of distributing the manufactures.[4] The industry tends, in general, to be oriented increasingly towards the market rather than towards raw materials.[5]

A rough and very arbitrary classification of the industrial sites would be those located (*a*) near the source of ore; (*b*) near the source of fuel; (*c*) close to the market, and (*d*) at some intermediate point. Examples are numerous in each of these categories.[6] The Corby, Wellingborough and Kettering works on the English Jurassic ores and Barrow close to the Cumberland haematite; the Lorraine, Luxembourg and Lower Saxon industries; the works at Duluth and Sault Ste Marie: all these illustrate category (*a*).

Works situated close to the source of fuel are no less conspicuous: Ebbw Vale, the Ruhr and Belgian industries, the Newcastle works in New South Wales, and the Pittsburgh works in the United States.

In a few rare instances the fuel and ore occur so close to one another that the plant may be said to be located on *both*. Birmingham, Alabama, is perhaps the best-known example (*see* page 133), but the nineteenth-century smelting industry that was based on coal-measures iron-ore in the Ruhr, in Poland and in some parts of England and Scotland must also be considered under this head.

Iron and steel industries, established near the sources of ore and fuel, have in many instances attracted consuming industries to them. In general the modern industry tends to be attracted towards its markets, but is restricted by the difficulty of finding suitable sites. The older works within the limits of great cities are one by one being squeezed out, and new works are being established as a general rule in the open countryside. The case of Chicago has already been mentioned. At Pittsburgh the industry has tended to move from the heart of the city to its distant periphery. The same is true of the Sheffield region and also in some degree of the Ruhr. There is no integrated works on Tyneside or the tidal Clyde, though these areas constitute very large markets for the products of the steel-mill.

Only in exceptional circumstances is an iron and steel works today likely to be established on or very near an ore deposit. If we may assume that a modern works will be planned to produce at least one million tons of steel a year (*see* below, page 66), and that the works will have a useful life of at least twenty-five years, then an ore mass of some 50 million tons, depending on its grade, must be assured.[7] A smaller deposit could not wisely be used in this way. There are not many such large deposits, and the number which are suited in other ways, without the admixture of other ores, to support a local smelting industry is small.

The chief sources of coking coal (*see* below, page 62) are less restricted, but a coal-mining region has little to recommend it as the site of heavy industry. The most recent extensions of the smelting industry in both the Ruhr and Upper Silesian regions have in fact been outside the limits of the exploited coal basin.

An intermediate location seems then to be the most desirable. The use of water-transport for ore is not only the cheapest but also the most flexible. A works may have been established beside a navigable inland waterway or on the coast usually for the purpose of taking ore from a specific source, but it can easily turn to other sources as may seem desirable. At the coast, furthermore, the ore has to be unloaded, and to stockpile it along the quays and to erect the furnaces directly behind them will save at least one handling of the ore.

After the invention of the Bessemer process the demand for low-phosphorus ores led in France and Germany, which were poor in such deposits, to the establishment of coastal works, able to draw this ore from whatever source was available.[8] In the United States the opening up of the Lake Superior ores was followed by the establishment of smelting and steel works on the shores of the Lower Great Lakes. Today the same works are preparing to use, along with the diminishing ore reserves of Minnesota, the output of the newer mines of Labrador.

The newest and biggest American iron and steel developments have been on the eastern seaboard. The Bethlehem Steel Corporation built up the Sparrow's Point works, near Baltimore, on the basis of imported ore which was unloaded directly on to its own stockpiles. Chilean ore, then Swedish and now Venezuelan and Liberian, have figured prominently in the furnace. As long as the ore can come by ore freighter, the Sparrow's Point works are well placed. More recently the United States Steel Corporation has established the Fairless works on the tidal Delaware river for the purpose of using Venezuelan ore.

It is impossible to evaluate separately the importance of an overseas source of ore and of proximity to the market of north-eastern United States, in the choice of the northern Atlantic seaboard as the site of an integrated works. The Fairless works have both these advantages and the erection of other works in the same geographical region is not impossible.[9] It has even been suggested that a works at Mobile, on the Gulf Coast, using high-grade ore imported by sea, would be more profitable than the works at Birmingham with their often-quoted advantages.[10]

In Great Britain coastal sites, or sites that can be reached without difficulty by sea-going ore boats, may be regarded as normal. The erection in the 1930's of a new integrated works at Ebbw Vale aroused strong comment; the site was distant from the coast and even farther from domestic sources of ore. But after the Second World War a large-scale works was built at Margam, on the Glamorgan coast, and a second integrated works is now being built on the coast—at Newport, Monmouthshire.

In France and Germany there seems also to be a revival in the significance of coastal locations. Developments are planned near Dunkirk and at the German North Sea ports.

Factors quite unrelated to transport costs sometimes influence the choice of site. A powerful factor in the choice of Ebbw Vale was the depressed nature of the region, and for similar reasons Jarrow, on the Tyne, was at one time considered as a possible site for an iron and steel works. Such social factors have never been significant in locating the American industry, but the choice of Nowa Huta, near Kraków, for a large, integrated Polish works was strongly influenced by the social problems in this poor and over-populated region.

The possession of a large, integrated iron and steel works not only gives economic strength to the country which has it but is thought also to confer political power and prestige. In under-developed countries the creation of such a works, however small the national contribution to it may have been, focuses the attention of the people. It gives a sense of independence from, almost of equality with, other and more developed nations. 'The whole nation builds Nowa Huta' was a slogan in Poland for several years, and it was not without its psychological importance. Others of the smaller and less-developed nations have derived a similar satisfaction from the completion of an iron and steel works. Whether, however, this sense of well-being has been achieved at too high a cost is discussed later in this book.

The Problem of Fuel: The problem of fuel supply has been present ever since in the later Middle Ages the forests began to be depleted by the charcoal-burners. We have already seen in Chapter 1 how mineral fuel gradually replaced charcoal as a fuel for the blast-furnaces in most countries. That the age of the charcoal-burning furnace is not yet completely over is shown by the recent erection of small furnaces to use this fuel in parts of northern Russia (*see* page 156). The coke-burning furnace, however, has now been dominant for a century, and all large modern works presuppose a supply of coke or of coking coal.

Coal consists, in addition to carbon, of ash, water and certain volatile constituents. These latter vary from 3 to 8 per cent in anthracite to 40 in long-flame and gas coals. Much controversy has gathered around the problem of the origin of these differences, but what concerns us here is the capacity of the different types of coal to yield coke suitable for use in the blast-furnace.[11]

Two qualities alone distinguish a good blast-furnace fuel. It should consist as nearly as possible of pure carbon. As much as possible of the sulphur, originally present in the coal, should have been removed, and the smaller the quantity of ash the better, as its melting and removal in the furnace consumes fuel and adds to the volume of the slag. Secondly, the coke should be firm and strong enough to withstand the immense weight of the furnace charge that is introduced on top of it.

Most metallurgical coke is made today in by-product ovens.[12] A battery of long, high but very narrow furnaces is charged with finely crushed coal and heated by burning gases which circulate around each oven. There are several patterns of furnace, but in all the gases given off by the coal as it is converted to coke are drawn off and used for heating after their by-products have been extracted. During the process the coke swells; the small pieces coalesce and, when they are ejected from the oven, form large, firm lumps of coke.

The types of coal capable of yielding a good metallurgical coke are limited.[13] In general they have only 20 to 30 per cent of volatile matter. Coals lying outside this range may yield only a soft and friable coke, incapable of supporting the weight of the furnace charge without crushing and obstructing the draught. Some types of coal, that of the Donetz Basin for example, are so sulphurous that too much sulphur is left in the coke for it to be a really good furnace fuel.

The coal-fields capable of yielding a high-quality coking coal are few indeed. In Great Britain, the Durham coal-field has always been considered outstanding, and the South Wales and Yorkshire coal-fields are good. In continental Europe, only the Ruhr coal-field produces a coking coal of unquestioned excellence. The Ostrava-Karvinná field of Czechoslovakia and the Central Belgian coal-field yield coking coal, but the coal-fields of France and the Saar are, by and large, unsuitable. In the United States, the Connellsville coal of western Pennsylvania is highly favourable, but most of the other large coal-fields of the world produce only coals that are unsuitable or only moderately suited for making coke. Under present technological conditions, India could not supply fuel for a really large smelting industry and South America is almost devoid of coking coal.[14] Indeed, it may be said that amongst the major iron-ore producers, only the United States and the Soviet Union have reserves of coking coal commensurate with their deposits of iron-ore. At the present time France, Luxembourg, Austria and the South American countries are heavily dependent upon imported metallurgical fuel, and in several others expansion of the smelting industry is possible only at the expense of importing fuel.

The seriousness of the fuel situation in some countries leads to experiments in other methods of smelting iron-ore. These have achieved some successs, though the blast-furnace remains and is likely to remain for a long time the most serviceable and economical instrument for separating iron from its ore.

Among the simpler modifications of customary blast-furnace practice has been the construction of low-shaft furnaces, in which the crushing weight of the charge is less than in furnaces of more usual design. The blast-furnaces in the Saarland and in Poland have long been kept far below the average size for Europe because the coke available was too soft for use in a higher furnace. In recent years the Swedes and the Japanese have used low-shaft furnaces, and experiments are now being made with them in East Germany. The current expansion of iron production in China appears to be achieved in part by this means. In Norway, Sweden and Switzerland electric smelting furnaces are in use. Some coke is still required for the chemical changes that take place in the furnace, but the heat is supplied by an electric arc.

A number of low-temperature processes have been introduced in recent years. Their general purpose is to separate the metal from the ore without completely melting it, as happened on the ancient forge. A 'sponge' iron results. Although it has certain disadvantages, it can nevertheless be refined to make steel, and the process can use almost any type of mineral fuel. The Krupp-Renn process is of this kind. Not only is it very adaptable in terms of fuel, but the basic unit is relatively small in scale and cheap to construct. It is well suited to those underdeveloped countries (*see* Chapter 8) whose requirements and resources do not justify the construction of an integrated blast-furnace works.

While, on the one hand, experiment aims at developing smelting techniques that are tolerant of soft or inferior fuels, attempts are being made to improve the quality of the traditional blast-furnace fuel. The chemistry of coal is a field that has made great progress in recent years, and experiments in blending coals from different sources have met with very considerable success. Most work has been done in this direction in France and Poland, countries richly endowed with coal but poor in the coking qualities. By careful treatment and blending with coal imported from Germany, the soft and volatile coal of the Saar and Lorraine can now be made to yield a serviceable coke. One may confidently expect that more progress will be made in this direction, and that the present imbalance between coking coal and iron-ore reserves will be in some measure rectified.

Not unrelated to this question is the progress that has been made during the last century and a half in economizing in the use of fuel. The ratio of coke used to iron smelted has steadily improved. The design of the furnace, the temperature and pressure of the blast and, more recently, the beneficiation and sintering of the ore and the maintenance of a high top pressure in the furnace, all contribute to the efficiency of the furnace. Although the ratio of fuel to iron will today vary from one furnace to another, and even in the same furnace from one time to another, it is usually somewhat less than one ton of coke to each ton of pig-iron. If a good coking coal yields about 70 per cent coke, then a ton of pig-iron requires about 1·4 tons of coal. In other words, a good-quality ore has very roughly to be matched by a similar quantity of coking coal.

So far we have been concerned only with blast-furnace fuel. But this is only one of many points in the operation of an integrated works at which fuel is consumed. The blast is heated. The coke-ovens, mixer and soaking-pits are fired. Immense quantities of fuel are burned in the steel furnaces, and power is required to operate not only the rolling-mill but also the whole elaborate mechanism of transport within the plant. In most modern plants the coke is produced in the works and the coal-gas, after removal of by-products, is used to fire the coke-ovens, heat the blast or generate power. For such purposes it can be supplemented by the gas discharged from the blast-furnace, as this contains sufficient carbon monoxide to give it a considerable colorific value. As a general rule, such by-product gases are not nowadays allowed to come into contact with

C

the metal owing to the danger that they might impart sulphur or other impurity to the latter. The steel-furnaces, which formerly burned a mixture of coke-oven and producer gas, today most often use oil. This fuel is easily controlled, clean to operate and does not bring any injurious substances to the metal. Tar and, where it is available, natural gas are also used in the open-hearth.

The Problem of Scale: How big to build such a works is a problem of immense difficulty. On the one hand, there is to be considered the amount of capital that can be invested in the undertaking, the size of the market and the availability of raw materials; on the other hand, the size of plant which yields the greatest economies in the operation of the processes to be employed and in the production of the articles desired.

It is highly improbable that works will be established without an adequate raw material base to support them, and usually modern smelting and steel works are placed so as to be able to draw their fuel and ore from a variety of sources. In retrospect, however, it is easy to point to older works that could have been better sited and to some that were badly placed. But, unfortunately, it sometimes takes many years to reveal the weakness of a geographical location. Ironically, it is the carefully planned works of the Soviet sphere that most often demonstrate how easy it is to choose an unsuitable site.

The size of plant in smelting, steel-making and rolling has been steadily increasing for a century. The newest blast-furnaces have an annual capacity of almost half a million tons of pig-iron a year, and for most normal production needs this may be regarded as at present the optimum size. Such a furnace scarcely requires more labour than one a quarter of this size, and the higher temperature at which it can be run itself constitutes a considerable technological advantage. It is not practicable to operate a single blast-furnace, if it is integrated with steel production, because it has periodically to shut down for relining and maintenance work. Most modern works have at least two large stacks.

At the opposite end of the steel-making process, the rolling-mill shows even bigger economies with greater size. As was noted earlier, a continuous strip-mill has to be large. It is not likely to have a capacity of much less than a million tons a year, and such a mill, when all allowances are made for the wastage of metal on the one hand and the utilization of scrap in the steel-furnaces on the other, would demand two modern blast-furnaces to keep it supplied. Other types of rolling-mills would have a smaller optimum size. A plant which rolls sections and heavy plate may have a capacity of 400,000 tons a year, or less. But such heavy goods can be rolled only from very large steel ingots, which have first to be rolled down to blooms or slabs in a blooming- or slabbing-mill, which itself has a large capacity.

The steel works itself is built up of many small units: converters, open-hearth furnaces and electric furnaces. These are most efficient when they are large, but, compared with blast-furnace and rolling-mill, they are small-scale producers, and the only question is how many units should be established between blast-furnace and rolling-mill.

Plant for preparing iron-ore and for producing metallurgical coke can be more easily adjusted to the size of the furnace. The equipment for disposing of slag from the blast-furnaces and also from the steel works is dependent in part on the size of the undertaking, in part on the grade of the ore used. It presents no serious difficulties unless the source and grade of ore undergo a sudden change. At Krupp's Rheinhausen works, the wartime shift from high-grade Swedish to low-grade Lorraine ore placed so great a strain on the equipment for handling slag that one blast-furnace had temporarily to be closed down. But this and many similar adjustments of scale can be made without great difficulty. There remains the overriding problem of the size of blast-furnace and of rolling-mill and the adjustment of these to the needs and potentialities of the country concerned.

Of course, highly specialized qualities of iron, such as *spiegeleisen*, ferromanganese and Bessemer iron, must be produced on a smaller scale than that indicated above, because the market for them is restricted. Blast-furnaces which produce only foundry iron or castings, and thus have not to be integrated with steel- and rolling-mills, arc usually small. An exception, however, is the Pont-à-Mousson works in Lorraine, where the furnaces supply iron for a highly mechanized foundry works. Similarly, certain pressed and forged goods are produced on a comparatively small scale and, in this respect, the operating efficiency of a small works is but little less than that of a large.

It would seem, both from theoretical considerations and from the actual practice in newly established works, that an integrated iron and steel works should not have a pig-iron capacity very much less than one million tons. A steel capacity of 1·0 to 1·5 million tons, with a somewhat smaller rolling capacity, would not only give reasonable economies of production, but would also fit well with such a blast-furnace output.

A plant of this size is, however, too large for many countries; it represents a great capital investment and its output would be excessive for the domestic economy to absorb in a measurable period of time. Lower labour costs may also encourage the erection of smaller plants. Plants having a capacity of 500,000 tons of steel, and even less, are being planned and built, especially in some of the less-developed countries.

The Problem of Scrap Supply: Ore, fuel and limestone are not the only materials used by a modern works. Metal scrap is today so important a raw material of the steel-making process, and also to a limited degree of the iron-smelting, that it exercises an important influence on the site of a plant. Scrap-iron and -steel are used in three ways. They are introduced in small quantities into the blast-furnace, and have the effect of making the charge richer and of lowering the ratio of fuel consumed to iron produced. But scrap is likely to be used in this way only if it is unusually cheap and abundant—a very uncommon state of affairs—or if it becomes essential for the smooth operation of a works to enrich the charge.[15]

Of far greater importance is the use of scrap in the steel works. In the open-hearth, pig-iron is melted down and, with the help of a slag, its impurities are removed. The greater the volume of impurity, the thicker will be the layer of slag required, and, in consequence, the longer and more costly the refining process. Scrap, especially if it has been well selected, is already refined. It has only to be melted down and an addition of scrap will in consequence increase the output of the furnace and diminish its fuel consumption. In the electric furnace, which makes more extravagant demands on fuel than any other, only high-quality scrap is as a general rule employed. In the open-hearth the usual practice is to take scrap to the extent of 45 to 55 per cent of the charge. But this too creates difficulties. Whereas pig-iron from the blast-furnace can be run into the furnace in a few

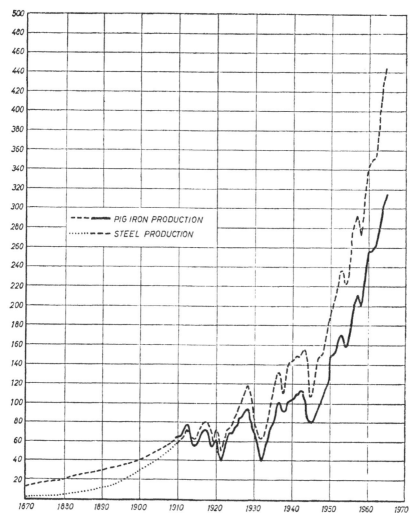

FIG. 4. Graph of iron and steel production 1870–1963 (based on
Statistics of the Iron and Steel Industries, British Iron and Steel
Federation). The increasing gap between iron and steel production
reflects the growing use of scrap metal

minutes, charging the scrap with the type of machine commonly employed may take hours.

The scrap is usually loaded into steel boxes which are picked up one by one and inserted through the doors of the furnace by the long arm of the charging machine. The lay-out of the steel works and the relative positions in it of the furnaces, the charging machines and the bay where the scrap-boxes are filled is important. It appears that the American plan leads to slightly more expeditious charging than that usually employed in Europe.

A modern integrated iron and steel works is planned on the assumption that from 40 to 60 per cent of the 'metallics' used will be scrap. If this proportion cannot for any reason be realized, then either part of the steel works, and consequently of the rolling- and finishing-mills, will be idle, or an excessive strain will be placed on the blast-furnaces. It should be noted here that scrap can be used only in very small quantities in the converter, and that the advantages of using scrap have contributed greatly to the modern extension of the use of the open-hearth.

Scrap comes from two sources. The more important is usually the works itself. The rolling-mill generates large quantities of scrap: the sheared ends of slabs and bars, the trimmings from the edges of the sheet, the large quantities of damaged and buckled pieces that arise even in the best-conducted works. The farther the works carries the finishing process, the greater will be the amount of scrap produced. This is 'home' scrap, and it can usually be assumed that in volume it will bear a nearly constant relationship to the amount of crude steel and finished goods made.

In an integrated works, however, it is unlikely to be sufficient, and must be supplemented by 'bought' scrap. This comes usually from a dealer. When old buildings and plant are dismantled, when automobiles and locomotives cease to be serviceable, when ships are broken up, scrap is produced. Tin cans and other household waste yield scrap. Old and broken farm equipment is an important source, and railway lines that have become mis-shapen with long use also go back, through the scrap-yard, to the steel works.

The volume of 'bought' scrap available varies with the level of 'civilization'. It is high in the United States; low in India and China, where the level of consumption of metal goods is very low. It varies also with the efficiency of the scrap-collecting agencies. Large pieces of scrap are rarely allowed to corrode away, but the small are sometimes left as unsightly marks upon the landscape. It is claimed that in Great Britain the scrap wastage is very slight. The small dealers or pedlars scour the countryside to sell to the larger, who sort and grade the material, and pass it on to the steel works. It has, indeed, been claimed by one of Great Britain's leading scrap-dealers that not more than 5 per cent of all the steel that is made ever fails to come back somewhere to the steel works. This recovery rate is certainly not attained in the United States, where the great distances and high wage rates combine to prevent recovery of such a large fraction. A recent report indicated that only some 10 per cent of the steel made will in fact be irrecoverable.[16] In China the intensity of the scrap-drive is such that only that which corrodes is lost.

A third and relatively unimportant use of scrap is in the foundry. Only cast-iron can, of course, be used, and the primary phase of sorting in the scrap-yard is the separation of iron from steel. It is probable that a higher proportion of iron comes back as scrap than of steel, because the more important losses—sunken ships, war losses and structural metal that is embedded in concrete—are of steel. It is true that the iron foundry, which uses scrap and new pig-iron, usually operates with a higher proportion of scrap than does the steel-furnace.

The problem of scrap supply is of importance in the context of this book for two reasons. It helps to locate a new works, and it has a significant bearing on the expected length of life of our iron-ore deposits.

Most steel works themselves provide about a third of the metallics they use from their own process or 'home' scrap. This scrap represents only a book-keeping transaction within the works. It is supplemented by purchasing up to a fifth of the total metallics in the open scrap market. Any large city, especially one with engineering industries, can produce a large quantity of scrap. The Fairless works, for example, is said to obtain all its purchased scrap from the vicinity of Philadelphia. Scrap availability, combined with market advantages, make any great city or group of cities a suitable site for a steel works.[17]

If we may carry this argument one step further, we find a very considerable generation of scrap in, say, the New York area, where there are very large scrap-yards, in Boston, in Mobile, New Orleans, San Francisco and Seattle.[18] The demand for scrap is not large in any of these cities, and, the relative costs of land and water transport being what they are, it pays to export scrap. In this way other countries—more particularly Japan and Italy, but also Denmark, Switzerland and Finland—are able to import scrap on a scale which allows them to reduce very greatly or even eliminate their production of pig-iron.

But the world market for steel is, despite its fluctuations, expanding rapidly.[19] There is thus a severe shortage of scrap, and the ratio of pig-iron to scrap in the steel-furnace is tending to rise. The graph Fig. 3 shows a widening gap between world production of pig-iron and of steel, a gap that can only be filled by the use of scrap. This state of affairs is difficult for industrialized countries like those of Western Europe and North America. It is very much more so for the relatively unindustrialized societies of Asia and South America. Although the steel works in these countries yield their quota of process scrap, it is difficult to supplement this with purchased scrap from the domestic market, because the consumption of steel per head has always been infinitesimal. Such countries have therefore either to import scrap and pig-iron or to erect additional blast-furnace capacity. The market for scrap has been and is likely to remain very tight, at least when business conditions are good. On the other hand, a blast-furnace works represents a very much larger capital investment than the steel works, and investment capital is just what the underdeveloped countries chiefly lack.

[1] It is probable that the best survey of iron and steel technology today is *The Making, Shaping and Treating of Steel*, 7th Edition, Pittsburgh, 1957, published by the United States Steel Corporation. A much shorter but very readable account is John Deardon, *Iron and Steel Today*, Oxford University Press, 1956. There is a good short account of the processes in W. Alexander and A. Street, *Metals in the Service of Man*, Pelican Books, 1944. Labour problems are briefly dealt with in *Men, Steel and Technical Change*, Department of Scientific and Industrial Research, H.M.S.O., 1957.

[2] See *The European Steel Industry and the Wide-Strip Mill*, U.N., 1953, for discussion of types of rolling-mill. See also *The Making, Shaping and Treating of Steel*, U.S. Steel Corporation, Pittsburgh.

[3] *World Iron Ore Resources and their Utilization*, U.N., 1950, 24.

[4] There are exceptions; *see* Allan Rodgers, in *Geographical Review*, XLII, 1952, 55–66.

[5] R. Hartshorne, 'Location Factors in the Iron and Steel Industry', *Economic Georgraphy*, IV, 1928, 241–53; Wilfred Smith, 'The Location of Industry', *Transactions and Papers, The Institute of British Geographers*, 1955, 1–18; Walter Isard, 'Some Locational Factors in the Iron and Steel Industry since the Early Nineteenth Century', *The Journal of Political Economy*, LVI, 1948, 203–17.

[6] For a listing of iron and steel works *see* H. G. Cordero, *Iron and Steel Works of the World*, 2nd Edition, London, 1957.

[7] See *World Iron Ore Resources and their Utilization*, U.N., Department of Economic Affairs, 1950.

[8] N. J. G. Pounds, 'Historical Geography of the Iron and Steel Industry of France', *Annals of the Association of American Geographers*, XLVII, 1954, 3–14.

[9] W. Isard and John H. Cumberland, 'New England as a Possible Location for an Integrated Iron and Steel Works', *Economic Geography*, XXVI, 1950, 245–59.

[10] H. H. Chapman, *et al.*, *The Iron and Steel Industries of the South*, University of Alabama, 1953, 369.

[11] F. J. North, *Coal, and the Coalfields of Wales*, Cardiff, 1931.

[12] R. A. Mott, *The History of Coke Making and of The Coke Oven Managers' Association*, Cambridge (Heffer & Sons), 1936.

[13] See *Analyses of Foreign Coals*, Bulletin 512, U.S. Bureau of Mines, 1952.

[14] *World Iron Ore Resources and their Utilization*, U.N., 1950, 15–17.

[15] Edwin C. Barringer, *The Story of Scrap*, Institute of Scrap Iron and Steel, Washington, D.C., 1954; 'Use of Scrap in Steel Making', *Iron and Coal Trades Review*, CLXXIV, 1954, 1271–2. This article is a summary of one which appeared in *Stahl und Eisen*, and contains some useful statistics.

[16] 'Final Report on a Survey and Analysis of the Supply and Availability of Obsolete Iron and Steel Scrap', Battelle Memorial Institute, Columbus, Ohio, 1957, in *Report on Iron and Steel Scrap by the Department of Commerce Pursuant to Public Law 631, 84th Congress*.

[17] W. Izzard and J. H. Cumberland, *op. cit.*

[18] See *The Minerals Yearbook*, U.S. Department of the Interior. A section on 'Iron and Steel Scrap' has been included since 1939.

[19] See *General Report on the Activities of the Community*, European Coal and Steel Community, published annually.

4

UNITED KINGDOM

IT IS in the United Kingdom, scene of so many of the significant advances in the metallurgy of iron and steel, that their impact on the geography of the industry can best be studied. An earlier chapter has reviewed briefly the reserves of iron-ore available to the British industry. These consist, in short, of several small deposits of good-quality haematite, to be found in the older rocks of the north-west; the coal-measures ores, encountered in the course of mining for coal, and the very much more extensive bedded ores of Secondary geological date. The latter are not limited to the familiar and important deposits of the scarplands of Yorkshire and the East Midlands. They occur also in the Wealden district of south-eastern England as well as in Wiltshire and Somerset. Great Britain is today one of the more richly endowed of European countries, though these reserves are made up largely of the low-grade, highly phosphoric ores of the oolitic belt.[1]

Whatever may be the economic and technical problems of mining, Great Britain's coal resources are large, and contain a high proportion of coking coal. Furthermore, they are widely scattered. Not all the British coal-fields produce coal of metallurgical quality, but this type of coal is obtained in particular from the Durham, Yorkshire and South Wales coal-fields. That of the Durham coal-field has probably enjoyed the highest reputation.[2] The ancillary materials—limestone, dolomite, fire-clay—are available at, in general, no great distance from the works, and the supply of iron and steel scrap and of water for cooling purposes is at least adequate. Thus, on a broad view, Great Britain may be said to have natural advantages at least as good as those of the other industrial nations of the world.

The Charcoal-iron Industry:[3] The iron industry of Great Britain antedates the Roman invasion by several centuries, but until the end of the Middle Ages it was practised on only a very small scale. Early in the sixteenth century the blast-furnace was introduced from the continent of Europe and speedily adopted in the Wealden district. Until the mid-eighteenth century the Weald of Kent and Sussex remained the most significant centre of iron-working. But the problem of fuel supply always presented difficulties. The Wealden forests were not extensive, and in general the requirements of the shipbuilders took precedence over the needs of the iron-worker. Furthermore, the small streams of this area were too small and too irregular in their flow to provide power throughout the year for the bellows and hammers.

The Forest of Dean, with reserves of haematite ore and wood-lands to which the shipbuilders had made little claim, succeeded to the position of the Weald.

But iron-smelting and -refining were widely carried on in the Welsh Marches, especially in Shropshire, in the West Midlands and along the eastern flank of the Pennines. An outlier of the industry lay in the Furness district of northern Lancashire. In all these areas both ore and charcoal were available. Ore was found more widely than charcoal, and it might be said that smelting was concentrated in those ore-bearing areas which still could provide scope for char-coal-burning. The Forest of Dean[4] and the Welsh Border produced pig-iron beyond local needs, and regularly sent it to the Birmingham area where it was refined and used.[5]

Coke-smelting and Puddling: Throughout the three centuries following the introduction of the blast-furnace into Great Britain the forest resources were slowly being depleted, and attempts to restore them were local and of little importance. The remedy lay of course in the use of coke in the blast-furnace. Early experiments in this direction were uniformly unsuccessful (page 17), and even after the efforts of Abraham Darby were in 1709 crowned with success,[6] the new process spread only very slowly. There were several reasons for this: the quality of the coal available for use, the technical capacities of the furnaces already existing, and the long-continued prejudice in favour of charcoal-smelted iron. When, however, the use of coke instead of charcoal did begin to achieve widespread acceptance, it brought with it extensive changes in the location of the industry. Charcoal-smelting had not disappeared a century after Darby had begun to use coke, but it had been reduced to insignificant proportions.[7]

The effect of using coke in the blast-furnaces was to draw the smelting industry to the coal-fields. That this did not also entail a large-scale movement of ore from existing mines to the newly established furnaces was due to the discovery and use of coal-measures ore. The Weald, the Forest of Dean and several outlying centres disappeared from the list of iron-producers. The Shropshire and South Yorkshire centres of the charcoal-smelting industry were so close to the source of coking coal that, although furnaces were replaced, the broad location of the industry continued. At the beginning of the nineteenth century the chief centres of the smelting industry were:[8]

Staffordshire and Shropshire	..	104,426 tons
South Wales	..	75,601 „
South Yorkshire	..	26,671 „
Central Scotland	..	23,240 „
Other areas	..	28,268 ,

This conspicuous trend towards the coal-fields was assisted by the introduction and spread of the puddling process. This method of making bar-iron from pig-iron has been described on page 19. The failure of Cort to substantiate his claim to a patent for the process merely increased the rapidity with which it spread. It was extravagant of coal, though not particular as to its quality, and the puddling process was drawn to an even greater degree than smelting towards the coal-fields. The Yorkshire coal-field with its adjoining areas in Derbyshire and Nottinghamshire, the Birmingham region, South Wales and Central Scotland became not merely the chief but the only centres of puddling. The making of steel by cementation and crucible processes (*see* pages 20–21) also required large quantities of fuel and it too came to be located mainly in Yorkshire and in the Black Country, around Birmingham and Wolverhampton.

The geographical pattern which had thus developed by the beginning of the nineteenth century was changed only in detail during the ensuing half-century. Furnaces multiplied in the coal-mining valleys of South Wales; the Black Country began to acquire its proverbial colouration, and from Sheffield the furnaces and steel works began to spread down the valley of the Don towards the Humber. New technology and new sources of ore then brought about a change in the relative importance of these and also of other areas.

New Ores and New Technology: The production of pig-iron expanded steadily during the first half of the nineteenth century. From less than a quarter of a million tons yearly at the beginning, it rose to over three million tons by the middle of the century.[9] A substantial part of this output had been achieved by using coal-measures ore. But the supply of this ore, so excellently located with reference to fuel, was beginning to fail. Thus, with the expectation of a continued increase in demand for iron, other sources of ore were examined. Ore began to be imported, especially from Spain, and a wider search was made within Great Britain.

Developments in technology were closely interwoven with the use of newly discovered ores. In 1851 the ore-bearing beds of Northamptonshire were rediscovered, and at about the same time John Vaughan first opened up the ores of the Cleveland Hills of North Yorkshire. The Cleveland ores were geographically the better placed. Only a few miles on the other side of the Tees lay the good coking coal of the Durham coal-field, and it was no serious transport problem to bring ore and coal together along the banks of the Tees in the neighbourhood of Middlesbrough. In 1851 the first furnace of Bolckow and Vaughan was blown in. The Bolckow-Vaughan works were followed by those of Lothian Bell, Dorman and Long and of several smaller and less illustrious firms. If one discounts the now scanty reserves of coal-measures ore, there was no place in Great Britain where ore and coal occurred in such close juxtaposition as here in North Yorkshire and County Durham. But the ore was low-grade and highly phosphoric. It could not be used to make iron for the Bessemer process, invented only a few years after the Cleveland ores had been discovered. Instead, iron was puddled, and as a number of branches of industry, including shipbuilding, clung obstinately to puddled iron for many years, there was no lack of outlet for iron refined by the older process. Nevertheless, Dorman and Long did establish Bessemer converters and used in them pig-iron smelted from non-phosphoric Spanish ores.[10]

Steel was, however, gaining in relative importance and it was the discovery of the Thomas process that permitted the Middlesbrough industry to go over completely to steel-making. Indeed, it was at Bolckow and Vaughan's works that Gilchrist Thomas's first successful 'blow' was made. Even before this, the iron industry of Middlesbrough had been growing faster than that of other iron-producing regions. Within twenty years of the establishment of the Bolckow and Vaughan works, the area was producing over a quarter of the pig-iron smelted in Great Britain, and had outstripped its nearest rivals, South Wales and Central Scotland.

The ores of Northamptonshire were rediscovered—for there had been some ancient and medieval iron-working here—at the same time as the Cleveland ores were opened up. But here in the East Midlands there was no such spectacular industrial development.[11] The production of ore was expanded more slowly than in the Cleveland Hills and the first blast-furnace, at Finedon, near Welling-borough, was not built until 1866. Even then the smelting industry grew slowly. There were several reasons for this. The Northampton-shire ores lay much farther from the coal-fields than did the Cleveland ores, and the railway network was not altogether favourable. Ore was shipped to the iron works of the West Midlands and Derby-shire, and the railway trucks brought back fuel for furnaces in Northamptonshire. But in these instances the ore was being supplied to long-established smelting works which showed little inclination to shift their operations to Northamptonshire. There were technical difficulties in the way of using the ore in furnaces designed to smelt other and different ores. Workmen disliked it and consumers were prejudiced against iron made from it. Lastly, the East Midland industry never in the nineteenth century fell under the control of men as competent technically and as aggressive commercially as those who dominated the industry of the north-east coast.

The smelting industry was expanded in the 1870's. A few pud-dling works were established, but by and large the area shipped pig-iron to other areas where it was converted to wrought-iron or steel. Lack of fuel was, of course, an important reason why puddling did not become important here as it had done in Middlesbrough. Throughout the last quarter of the nineteenth century there was little expansion either of iron-ore mining or of smelting. This was a period when the iron industry was relatively depressed, but the East Midlands scarcely shared in such growth as did take place.

Ores similar in geological origin and mode of occurrence to those of Northamptonshire occur also in the limestone ridge of Lincolnshire. Their existence was discovered in 1850,[12] and within a short time ore was being shipped to the iron works in South Yorkshire and Derbyshire. The smelting industry, established here in 1865, at first produced only iron for the foundry and puddling-furnace, but in 1890 began to smelt iron suitable for the basic process. From that time steel-making began to grow in this area, though this remained until after the First World War a comparatively small producer. The start which the Scunthorpe area thus had over the East Midlands in the manufacture of steel is probably to be attributed to its closer associations, both geographical and commercial, with Yorkshire. The coal-mines around Doncaster lay only twenty miles from Scunthorpe, and today the largest of the Scunthorpe works, the United Steel Companies, embraces within its control some of the larger steel-making and steel-using firms of the Sheffield area. An exchange of coal and metal could be organized with greater ease between Scunthorpe and the Sheffield region than between Northamptonshire and the West Midlands.

Retrogression and Change: The growth in pig-iron and steel production in the vicinity of the Jurassic ores was reflected in the relative decline of the industry in its older centres of production. In South Wales the production of pig-iron remained almost static throughout the half-century before the First World War, and came to depend before the end of this period almost entirely on imported ores. This in turn brought about changes in the location of works within the area. In the West Midland region Shropshire ceased to smelt pig-iron, and in the Black Country itself there was on balance a small decline in smelting. The Sheffield district—a large consumer of pig-iron rather than an important producer—the North Lancashire and Cumberland area, where there were good haematite ores, and Central Scotland maintained their position through this half-century, but only the north-east coast and the scarplands of Northamptonshire and Lincolnshire showed any large gains.

It is beyond the scope of this chapter to consider the stagnation of the iron and steel industry in the later years of the nineteenth century. Its reasons were basically non-geographical. After the vigour which the industry had displayed during the first two-thirds of the century, a lethargy settled upon it. Individual firms showed little desire to expand either by internal growth or by amalgamation. Plant, compared with German and American, was uneconomically small, and integration of the several facets of the industry made less progress than in Great Britain's principal rivals.[13]

This state of affairs was not wholly remedied during the inter-war years. There were amalgamations of firms in the industry and, consequent upon them, plant was reorganized. Notable amongst such changes was the fusion of Dorman and Long and Bolckow and Vaughan and the rationalization of the Middlesbrough industry which followed. In South Wales four leading steel-making firms, including Baldwins and Richard Thomas, were merged; the United Steel Companies was formed from works at Sheffield, in Lincolnshire and in Cumberland, and there were other structural changes in the industry, the general effect of which was to create larger and more effective units.[14]

A number of large modern works was established; John Summers erected a new works at Shotton on the river Dee, the United Steel Companies put up the Appleby-Frodingham works in Lincolnshire and Lysaght's, the nearby Normanby Park works. Then in 1936 came the completion of the Great Corby works in Northamptonshire, of Stewart and Lloyd's. These locations were wise. If low-grade ores were to be used, then an ore-field site was the most desirable in view of the low ratio of fuel to ore achieved in a modern furnace. Then came the decision of Richard Thomas to build a large integrated plant, with continuous strip-mill, at Ebbw Vale, a site nowhere near ore and indeed not easy of access. Ebbw Vale was chosen in preference to Redbourne in Lincolnshire under pressure from the government, which acted from social rather than economic motives. It rescued a depressed area but thereby did nothing to help a depressed industry.[15]

Despite these changes, the British industry remained fragmented and badly located, at least in comparison with that of its most important continental rivals. Public policy was obsessed with the problem of the Depressed Areas, and its motives were, in the words of D. L. Burn, 'a curious blend of the compassionate and the commercial'.[16] It ignored the fact that changing location is of the essence of economic growth. The greatest good of the greatest number could best be achieved by the further concentration of iron and steel plant and their relocation in low-cost areas of production. Policy in this respect lacked the boldness which characterized both German and American, and in some instances the result was to impose a depressed industry upon an already Depressed Area. But the government's position was not an easy one. Labour was notoriously immobile; some areas clung with a proprietary instinct to their long-established industries. In such desirable locations as Clydeside there was just not enough room for a large integrated works, and elsewhere the conflicting claims of agriculture and public amenities made the choice of industrial sites difficult. It was not altogether that Americans and Germans had broader minds; they had broader acres over which their minds could experiment.

THE REGIONAL DISTRIBUTION
OF THE IRON AND STEEL INDUSTRY

The iron-smelting and steel-making processes have come, in the ways that we have already seen, to be concentrated in some half-dozen producing areas. These 'regions' are in part relict areas in which the industry has survived from earlier times; in part, modern planned developments in which full regard has been had of current production costs and sources of materials. Even within a single small industrial region, such as South Wales, plants range from the obsolete and the obsolescent to the most modern; from works that were well located with reference to raw material sources that have now been exhausted, to those laid out with regard to the likely sources of supply for the foreseeable future.

The British industry is characterized by a large number of operating units, only a few of which are modern, integrated works. Some are, of course, small and highly specialized steel producers; many are small and old. The industry is distinguished, secondly, by its high degree of dispersion. Works may, however, be grouped geographically according to whether they are oriented towards (a) the coal-field, like the Midlands and the West Riding of Yorkshire; (b) the iron-ore field, like great Corby and Scunthorpe, and (c) the coast, for the convenience of importing ore, such as the Teesmouth and the new plant in South Wales. Such a classification is arbitrary. The orientation of some areas has changed, as has the Teesmouth area, from dependence on local to dependence on imported ore.

South Wales: Most important of these regions of iron and steel production is South Wales, with an output in 1963 of 5,865,000 tons of steel or nearly a quarter of the British total. The South Wales industry is an interesting amalgam of the old and the new. The attraction of the region lay originally in the ironstone that could be got from the northern margin of the South Wales coal-field.[17] It was smelted, at first with charcoal, and during the later years of the eighteenth century a group of iron-smelting centres grew up in the mountain valleys between Blaenavon and Hirwaun. For a time South Wales was the most important producer of pig-iron in the British Isles. But the ore reserves, never large, began to run out in the middle years of the nineteenth century. One by one the works in this area closed, and the last furnaces were broken up early in the present century.[18]

But South Wales had an abundance of good coking coal as well as a local market in the tinplate industry.[19] The smelting and refining industry of South Wales declined relatively to the rest of the country and its location shifted. There had always been some iron works near, if not on, the coast. These increased in importance as the works in the interior declined.[20] Imported ore gradually replaced local ore, and a location near the coast, especially where, in western Glamorgan, the coal-field reaches the sea, had cost advantages over positions far inland. Smelting has thus come to be concentrated in works located on the tideway at Cardiff, Port Talbot and Briton Ferry. There are, however, a number of steel-making- and rolling-mills at inland sites, both in the tinplate region to the west and in the older centres of the iron industry in the eastern part of the coal-field. A very large proportion of the steel made in South Wales is rolled into thin sheet and is then tinned or galvanized. There is relatively little development of the engineering and shipbuilding industries.

In 1936, as has already been noted, a large, integrated works was established at Ebbw Vale, an area where the smelting and steel industry had been declining for a century. It brought employment

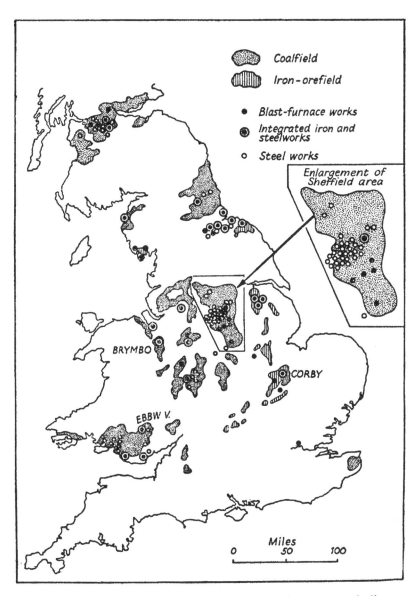

FIG. 5. Iron and steel works in the United Kingdom, excluding
Northern Ireland

to the area at the expense of shipping low-grade ore by rail a distance of 150 miles from Northamptonshire. Since the war, an even larger plant has been built as a part of the Steel Development Plan at Margam on the Glamorgan coast.[21] It is a large, integrated works, producing steel sheet for electrolytic tinning, by means of a continuous strip-mill. It uses South Wales coal and imported ore, which it receives through its own docks. These two developments have restored the South Wales region to the position of primacy which it lost with the rise of the Teesmouth industry. The new plant which is being built at Newport will add yet more to the capacity of South Wales.

Teesmouth: The rise of an important smelting and steel-making industry on Tees-side began with the recognition in the middle of the last century of the Cleveland ore deposits. The ore-mines lay a few miles south of the Tees, and only a little farther to the north of the river was the Durham coal-field, with some of the best coking coal in Great Britain. The Cleveland ore was phosphoric, but even before the introduction of the basic process this had become the most important centre of iron-smelting and puddling in the country. The Thomas process and later the basic open-hearth merely added to the advantages possessed by Teesmouth. Smelting and iron and steel works were built on both sides of the river as well as on the coast nearby.[22]

Towards the end of the nineteenth century it became apparent that the Cleveland ores were running out. By 1900, the Cleveland mines supplied only a little more than a half of the ore used. When the Second World War began, one-third of the metal content of the ores smelted was from Cleveland ores, and by 1954 this had sunk to 9 per cent. Most of the remainder was from foreign ores. This change has greatly affected the competitive position of the works in the Middlesbrough-Hartlepool area, though it retains of course the great advantage of the good Durham coking coal. Most of the works in the area have, by a series of mergers, passed into the possession of Dorman, Long and Co. This company conducted an extensive reorganization of plant between the two World Wars, and another rationalization and extension programme is now in progress. There are today three integrated works in or near Middlesbrough, together with some separate blast-furnace works and steel-mills. There are, furthermore, three integrated works on the coast at West Hartle-pool, Redcar and Skinningrove and one integrated works, common-ly associated with this group, at Consett, thirty miles away. The region as a whole concentrates on the production of heavy sheet and 'shapes', such as rails and structural steel, much of which is used in the shipbuilding industry. Owing to the proximity of the materials, there has always been a strong emphasis on smelting. The ratio of pig-iron to steel production is high, and the area nor-mally sends pig-iron to essentially steel-making regions such as the West Riding and the West Midlands.

The Black Country: The production of iron and steel in the West Midlands is a relict industry, surviving from the time when the local coal-measures ore, fuel and limestone alone sufficed to provide the materials.[23] Charcoal-iron was made here in the early seventeenth century, and here Dud Dudley claimed to have experimented with the use of coke. Iron-using industries, especially the manufacture of nails and agricultural tools, grew up here, and the local supply of pig-iron had to be supplemented by imports from the Welsh Border and South Wales. Coke-smelting began to replace charcoal in the second half of the eighteenth century, and the introduction of puddling further emphasized the importance of the coal-field of the Birmingham plateau. Although the smelting industry was ex-panded, there continued to be a net import of pig-iron into the area, and the puddling-furnace became the foundation of the West Mid-land iron industry. This industry reached the peak of its prosperity in the middle years of the nineteenth century. Thereafter, changing technology and the exhaustion of local materials led to the decline and, in very recent years, the disappearance of the puddling in-dustry. Local ores were exhausted and local coal was, by and large, not of coking quality. Lastly, the Bessemer and subsequent steel-making processes not only competed with and then replaced the Black Country's puddling industry, but focused attention on other areas better placed to practise them.

Iron-smelting is still practised on a small scale, and ore is brought into the area, mainly from the East Midlands. The furnaces produce special types of iron. Puddling has almost ceased, but the engineer-ing industries of the Birmingham area continue to be supplied, in part at least, with steel from the mills of the Wednesbury district.

This West Midland steel industry uses pig-iron imported from the north-east coast or the East Midlands, and also a high proportion of scrap, which is relatively abundant in this centre of the engineering industries.

West Riding of Yorkshire: The iron and steel industry of the Don valley and of neighbouring parts of Derbyshire is also in large measure a relict industry. It developed, like that of the West Midlands, on the basis of local fuel and coal-measures ore, and formerly extended northwards to Leeds.[24] But the exhaustion of the local ores has robbed this area of much of its advantage. The more northerly works have closed, but in the Rotherham and Chesterfield districts iron-smelting continues to be important on the basis of Lincolnshire ores. Much of the iron produced is of foundry quality, and Rotherham continues to make iron-castings. Sheffield itself is the most important centre in Great Britain for the manufacture of high-quality steel. Its iron-smelting industry has long since ceased, but already in the mid-eighteenth century steel-making in the hands of Benjamin Huntsman had begun to develop. Sheffield was the scene of many of Bessemer's experiments in steel-making, and as the converter and open-hearth processes spread more widely, Sheffield itself concentrated on the production of quality steels and more recently of alloy and high-speed steels, most of the British production of which is from here. The local engineering industries produce a relatively large proportion of scrap, which is supplemented by pig-iron from the East Midlands, Lincolnshire and other blast-furnace centres.

East Midlands and Lincolnshire: The opening up of the Jurassic ore deposits of the East Midlands and Lincolnshire has already been mentioned. The mining of the ore began in Northamptonshire in 1852 and in Lincolnshire seven years later. The first blast-furnace was built on the ore-field in North Lincolnshire in 1864, and in Northamptonshire in 1866. For a considerable period both regions smelted foundry-iron or pig-iron for steel manufacture elsewhere. Steel-making was first introduced in the Scunthorpe area in northern Lincolnshire, and here three large, integrated works have been established. In the East Midlands the Corby works were opened in 1936. These steel works produce plates, sections and tubes, concentrating in the main on heavy rolled goods. In the East Midlands three old blast-furnace works, which had survived from the nineteenth century, have recently been closed.

These iron and steel works on the iron-field are at present reckoned to be low-cost producers. It does not necessarily follow, however, that new plant should be located in these areas. There have been, as we have already seen, powerful forces pulling the industry towards the older if less economical centres. It has not been easy to attract labour to the ore-fields or to satisfy other interested bodies that this is a proper use for the limited areas of land available. Lastly, it is certain that the difficulty, and hence the cost, of extracting the local ores will rise and the relative profitability of the site decline.

Other Centres in England and Wales: A few other centres of iron-smelting and steel-making deserve mention, some of them survivals from an earlier age, others recent developments. Among the former are the iron-smelting works of the Furness district of Lancashire and the works farther north at Whitehaven. These all grew up to smelt the high-quality, local haematite. At first charcoal was used as fuel, then coke from the Durham coal-field and more recently fuel from the Cumberland coal-field. The short period which elapsed between the invention of the Bessemer process and the discovery of the basic lining was one of great prosperity and growth, because the haematite of this area was uniquely suited to the acid process. Workington continues to make steel by the acid Bessemer process, and the Furness district ships low-phosphorus pig-iron to the steel-making centres. But the haematite deposits are nearing exhaustion, and at present about three-quarters of the blast-furnace charge is made up of imported ores.

In South Lancashire and in nearby Cheshire and Flintshire are several plants as diverse in origin as they are in location. The old iron-smelting industry of South Lancashire has disappeared, and has been replaced by a new, integrated works at Irlam, on the Manchester Ship Canal. Ore is imported from overseas as well as to some extent from the scarplands ore-field. The Shotton works at the head of the Dee estuary and the Mostyn works, farther down the estuary on its western shore, are modern developments located near if not on a coal-field and easily accessible to overseas ore. The Brymbo works, near Wrexham, the Oakengates works in Shropshire and the Stoke-on-Trent works survive from the older, nineteenth-century pattern of industry. The small blast-furnace at Ford's Dagenham works is, on the other hand, of quite recent date.

Scotland: The Scottish smelting industry, like that of the older centres in England and Wales, grew up in the eighteenth century to use the blackband ores of the coal measures. Although there had been numerous small and ephemeral smelting works in the Highlands,[25] the first works to use coal and to employ the new methods was the Carron works, established near Falkirk in 1759.[26] Many works sprang up in the following years on the Lanarkshire coalfield. They received great encouragement when, in 1828, Nielson first used a hot blast. This not only allowed a more effective use to be made of the blackband ores, but also permitted the local 'splint' coal to be used directly in the blast-furnace without coking. Throughout the middle years of the nineteenth century the Scottish industry grew and flourished on the basis of blackband ore and 'splint' coal. The works were mainly in the Clyde valley, above Glasgow, with a secondary concentration in northern Ayrshire.

In Scotland, as in Yorkshire, the West Midlands and South Wales, the blackband ores were nearing exhaustion in the second half of the nineteenth century, and soon afterwards a shortage became apparent of 'splint'. The industry had its local basis cut away from under it. Many works closed, and the blast-furnaces that continued to operate were obliged to turn increasingly to imported ores. Even for that they were not well placed. None of the four blast-furnace works is accessible to large sea-going ore-carriers, though the Carron works are able to make use of the Grangemouth Canal.[27]

A consequence of this is that pig-iron production has declined in importance relative to steel production. Pig-iron is brought in from the smelting centres in northern England, and this region makes a proportionately greater use of scrap than any other in Great Britain. Steel-making- and rolling-mills are particularly numerous; there are today twenty-six separate works, of which Colville's Clyde Iron Works was until very recently the only one that was fully integrated. The total production of steel is about 2·1 million tons, so that some of these works are necessarily very small and specialized. Most lie between Glasgow, Coatbridge and Motherwell, near but not adjoining the blast-furnaces. The only integration that had been achieved before the war—at Clyde Iron and Clydebridge—necessitated carrying the hot metal across the river Clyde.[28]

The chief products of the Clyde steel-mills are predominantly heavy plates and sections, structural steel and tubes. A considerable amount—over 20 per cent—is used in the shipbuilding industry. It is significant that the whole of Glasgow lies between the steel works and the shipyards along the lower Clyde. Although the iron and steel industry is being extended, it is inconceivable that a new plant should be built on the Clyde below Glasgow: there is no room.

Despite the fact that the local area now yields scarcely any of the materials of the iron and steel industry, the importance of the market for steel in central Scotland remains a very powerful attraction to the industry. The more intensive competition for steel scrap of recent years has necessitated an extension of smelting and the high level of activity in the Clydeside shipyards has justified an expansion of smelting as well as steel-making and rolling capacity in this area. The most important event of recent years has been the building of the Ravenscraig works, near Motherwell, by Colvilles Ltd. This consists of a large blast-furnace works, integrated with the much older Dalzell plant, which is itself being modernized and extended. Only a steel works remains from the industry that was once widely spread in Ayrshire. But in the Clyde valley the impetus of history, tradition and an expanding local market have overcome the failure of local raw materials.

THE STEEL DEVELOPMENT PLAN

In the later years of the nineteenth century the British iron and steel industry ceased to expand at its earlier rate of growth and its competitive position began to weaken. Despite the stimulus of the First World War and tariff protection, iron and steel continued to decline during the inter-war years, relative to the rest of the British economy. To some extent this can be explained by the exhaustion of those resources on which the industry had been based in the nineteenth century. But national resources were very far from exhausted. The judgment of T. H. Burnham and G. O. Hoskins[29] is that 'there was no factor in raw material supplies that would account for the relative retrogression of some products'. The same authors comment upon 'the patch-work growth of British iron and steel works', their small scale, their scattered location, and the lack of planned integration from both the technical and the business angle. Capital investment was high relative to capacity, and, as a general rule, economies of scale could not be achieved.

That the seriousness of this situation was realized before the war was shown by the establishment of the new, integrated works at Great Corby and Ebbw Vale. But the location of the latter works in South Wales instead of in Lincolnshire, however justifiable from the social point of view, was economically a retrograde step.

In 1946 the First Steel Development Plan was published.[30] Its purpose was to increase steel capacity progressively to 16 million tons a year. This was to be done by the establishment of a large, new, integrated works in South Wales, and by enlarging the more favourably located of the older works. The original plan was continuously modified and extended, as the expected demand for steel ten or fifteen years ahead was seen to rise. The long-term objective of the industry is now 29 million tons of steel in 1962.[31]

Ore Supply: It was assumed that domestic ores would play an increasing role in the smelting industry, and that most of this would be from the East Midland and Lincolnshire fields. Here the ore can be obtained cheaply by strip-mining methods, at the price, however, of the destruction—temporary or permanent—of agricultural land values. There is not space here to discuss this much-publicized question. It is to be expected, however, that in time this mining will be pursued underground, as in Lorraine, with higher production costs, but without wide-scale injury to farmland.[32] The rate of increase in the smelting industry has been far greater than that of domestic ore production. Imports have increased and now constitute a very much higher proportion than the pre-war average.

			Home Ore	Imported Ore
			(million tons)	
1935–8 average	..	..	12·28	4·88
1946	..	..	..	12·75	5·57
1950	..	..	..	12·76	8·13
1956	..	..	..	16·39	12·77
1963	..	..	..	14·84	14·28

(*Development in the Iron and Steel Industry*, Table 13.)

The smelting industry has not expanded as had been expected, and ore consumption is no higher than in the late '50s. This large volume of imported ore, chiefly from Sweden and Norway, North and West Africa, Canada and Venezuela, further emphasizes the importance of coastal sites for smelting works.

The expansion both of the home production and of the import of ore has necessarily been slow. At the end of the war there was a shortage of shipping, and the solution of this problem has had to await the completion of a fleet of specialized ore-carriers. Docks have had to be adapted to the carriers, and rapid unloading machines built to expedite the turn-round of ships.[33] Where the works are not situated directly on the coast, special systems of rail transport have been devised to reduce as far as possible ore-haulage over the public railway system. The ports of entry for imported ore are Glasgow, the Tyne (for Consett), the Tees, Birkenhead (for Shotton) and the South Wales ports of Port Talbot, Margam, Cardiff and, in the future, Newport. In the size of ship and the scale of the unloading machines these ports are beginning to approximate to the level of American equipment.

Modernization of Plant: When the Steel Development Plan was put into effect, plants were mostly small and in many instances obsolete and uneconomically laid out. The Plan called for extensive modernization of, and additions to, plant that appeared to justify them and the eventual closing down of certain uneconomic works. The large number and the wide extent of these projects[34] shows that none of the existing iron- and steel-producing areas is to be eliminated from the field; the Brymbo works, for example, despite a disadvantageous location, are being modernized. The weight of the new plan, however, is being concentrated in two types of area: the English ore-fields and sites fairly well placed to receive imported ore.

In the words of a recent report of the Iron and Steel Board: 'An ideal steel works would be close to ore and coal supplies and near its market. If it were based on home ore it would be on the ore-field itself, and if it were based on imported ore it would be immediately beside a deepwater dock. In this sense there are few ideal sites for steel works and their location has generally to be a compromise.'[35] The new Margam works have poor dock facilities, but the Shotton works of John Summers Ltd. obtain ore through Birkenhead (fifteen miles), the Ravenscraig works of Colvilles Ltd. through the port of Glasgow. The large extensions to the Dorman Long plant at Middlesbrough, which are welding the Redcar, Lackenby and Cleveland works into a functional unit, have been influenced as much by the availability of unoccupied land as by the desirability of a waterside location for the smelting works. This country is too small and too crowded for the industry always to be able to establish itself on the economically right site.

Expansion of Iron and Steel: The success of the development plans is reflected in the expansion of iron and steel production. The rate of expansion of steel production was at first somewhat higher than that of pig-iron. This reflected the relative abundance of scrap in the post-war period. The expansion of blast-furnace capacity has been somewhat slower, but the ratio of pig-iron to scrap used is now increasing as new blast-furnaces come into production.[36] The following table illustrates the growth since 1936:

	Pig-iron	Steel
	(thousand tons)	
1936–8 average ..	7,781	11,910
1946	7,886	12,899
1947	7,910	12,929
1948	9,425	15,115
1949	9,651	15,803
1950	9,787	16,554
1952	10,900	16,681
1954	12,074	18,817
1955	12,670	20,108
1956	13,382	20,991
1957	14,512	22,047
1958	13,183	19,880
1959	12,784	20,571
1960	16,116	24,695
1961	14,984	22,441
1962	13,692	20,491
1963	14,591	22,520
1964	17,551	26,651
1965	17,740	27,439

(*Quarterly Bulletin of Steel Statistics*, United Nations (E.C.E.) and *Minerals Yearbook*, 1957.)

The plans of the past ten years have charted a new course for the British iron and steel industry. It is intended that plant shall be as large as possible within the framework of specialized national demands. The plants will be, wherever possible, fully integrated, with all departments from coking and sintering to steel-rolling closely placed for heat conservation and cheapness of transport between them. While the production of high-quality steel remains predominantly in its long-established centres of production, the large, integrated works which produce the standardized steel goods will be concentrated more in the lowest-cost producing areas. These are the Lincolnshire and East Midland ore-fields, and the coastal sites able to receive domestic coal and imported ore.

[1] *The Northampton Sand Ironstone: Stratigraphy, Structure and Relief*, Geological Survey, H.M.S.O., 1951.

[2] S. H. Beaver, 'Coke Manufacture in Great Britain: a Study in Industrial Geography', *Institute of British Geographers: Transactions and Papers*, 1951, 131–48.

[3] No short account of the British iron and steel industry can improve on those presented by S. H. Beaver in L. Dudley Stamp and Stanley H. Beaver, *The British Isles*, 4th Edition, London, 1954; and Wilfred Smith, *An Economic Geography of Great Britain*, London, 1948, Chapter 7. The history of the charcoal iron industry is exhaustively studied in H. R. Schubert, *History of the British Iron and Steel Industry*, London, 1957.

[4] F. T. Baber, 'The Historical Geography of the Iron Industry of the Forest of Dean', *Geography*, XXVII, 1942, 54–62.

[5] B. L. C. Johnson, 'The Charcoal Iron Industry in the Early Eighteenth Century', *Geographical Journal*, CXVII, 1951, 167–77.

[6] Arthur Raistrick, *A Dynasty of Iron Founders*, London, 1953.

[7] H. Scrivenor, *A Comprehensive History of the Iron Trade*, London, 1841, 83–97. The best account of the industry during the Industrial Revolution is T. S. Ashton, *Iron and Steel in the Industrial Revolution*, Manchester University Press, 1951. *See also* H. G. Roepke, 'Movements of the British Iron and Steel Industry—1720 to 1951', *Illinois Studies in the Social Sciences*, Vol. 36, Urbana, 1956.

[8] H. Scrivenor, *op. cit.*, 97.

[9] J. H. Clapham, *An Economic History of Modern Britain*, I, *The Early Railway Age*, Cambridge, 1939, 425 *et seq.*

[10] M. W. Flinn, 'British Steel and Spanish Ore', *The Economic History Review*, second series, VIII, 1955, 84–90.

[11] S. H. Beaver, 'The Development of the Northamptonshire Iron Industry, 1851–1930', in *London Essays in Geography*, edited L. Dudley Stamp and S. W. Wooldridge, London, 1951, 33–58.

[12] O. D. Kendall, 'Iron and Steel Industry of Scunthorpe', *Economic Geography*, XIV, 1938, 271–81.

[13] D. L. Burn, *The Economic History of Steelmaking 1867–1939*, Cambridge, 1940, especially 219 *et seq.* T. H. Burnham and G. O. Hoskins, *Iron and Steel in Britain 1870–1930*, London, 1943.

[14] T. H. Burnham and G. O. Hoskins, *op. cit.*, 208–10.

[15] D. L. Burn, *op. cit.*, 459–61.

[16] *Ibid.*, 513.

[17] A. H. John, *The Industrial Development of South Wales 1750–1850*, Cardiff, 1950; *ibid.*, 'Iron and Coal on a Glamorgan Estate', *Economic History Review*, XIII, 1943, 93–103.

[18] John P. Addis, *The Crawshay Dynasty*, Cardiff, 1957.

[19] W. E. Minchinton, *The British Tinplate Industry*, Oxford, 1957.

[20] H. C. Darby, 'Tinplate Migration in the Vale of Neath', *Geography* XV, 1929–30, 30–35.

[21] 'The South Wales Steel Industry', *Monthly Statistical Bulletin*, British Iron and Steel Federation, November, 1956.

[22] 'Steel on the North-East Coast', *Monthly Statistical Bulletin*, B.I. & S.F., July, 1955; 'Company Histories I; Dorman Long', *Steel Review*, April, 1957, 11–23.

[23] *Birmingham and its Regional Setting*, British Association for the Advancement of Science, Birmingham, 1950; especially 161–210; W. H. B. Court, *The Rise of the Midland Industries*, Oxford, 1953.

[24] H. Scrivenor, *op. cit.*, 181.

[25] Sir Ivison Macadam, 'Notes on the Ancient Iron Industry of Scotland', *Proceedings of the Society of Antiquaries of Scotland*, XXI, 1886–7, 89–131.

[26] Henry Hamilton, 'The Founding of Carron Ironworks', *The Scottish Historical Review*, XXV, 1928, 185–93; *ibid.*, *The Industrial Revolution in Scotland*, Oxford, 1932.

[27] 'Steel in Scotland', *Monthly Statistical Bulletin*, B.I.S.F., April, 1955.

[28] Anthony Rhodes, 'Company Histories 3; Colvilles', *Steel Review* January, 1958.

[29] *Op. cit.*, 135.

[30] *Iron and Steel Industry*, Cmd. 6811.

[31] *Development in the Iron and Steel Industry*, Iron and Steel Board Special Report, 1957, 3.

[32] A start has recently been made in Northamptonshire; *see* Stewarts and Lloyds Ltd. annual report, *The Times*, January 3, 1958.

[33] 'The Handling of Imported Ore', *Monthly Statistical Bulletin*, B.I.S.F., January, 1954; 'Foreign Ore', *Steel Review*, January, 1958, 16–38; *Development in the Iron and Steel Industry*, Iron and Steel Board Special Report, 1957, 44–45.

[34] Listed in Cmd. 6811, Iron and Steel Board Special Report, 1957, and *Annual Report of the Iron and Steel Board*, published annually.

[35] *Development in the Iron and Steel Industry*, 1957, 66.

[36] *Development in the Iron and Steel Industry*, 1957, 30–31.

5

THE IRON AND STEEL INDUSTRIES
OF WESTERN EUROPE

IRON-WORKING has been practised in Europe for some three thousand years. As in Great Britain, the geographical pattern of the industry has been continuously changing under the influence of developing technology and changing resources. During the first two thousand years of this period, iron works were few and their equipment slight. Though occasionally the slag from such a works or a corroded fragment of iron may be found, the evidence does not allow us to form any precise geographical picture of the industry. During the Middle Ages the pattern became clearer. Over the whole of Western Europe there was a thin scatter of iron works, but amongst them was a denser concentration in a few areas where resources were more abundant and skills more highly developed.[1]

The Early Modern Pattern of Industry: Even at this early date the Rhineland showed signs of unusual activity. Bloomeries were particularly common in the Ardennes and Eifel hills to the west of the river Rhine, and in the valleys of the Sieg, Lahn and Dill to the east, and, as we have already seen, the blast-furnace was probably evolved from the simpler bloomery in this region. The Harz mountains and the ranges of hills that border Bohemia all developed their iron industry, and the Austrian provinces of Styria and Carinthia may have experienced a more rapid expansion of iron-working than any of them. Central Italy, Savoy and Piedmont, the Pyrenees and the region of dissected limestone plateaus in Champagne and Burgundy, would all show closely clustered iron works against a background of more widely scattered units over the rest of Europe.

Reasons for this incipient concentration were varied, and were certainly not clearly known even to those who brought it about. Such areas had to have ores either in greater abundance or of superior quality to those found elsewhere. In his empirical fashion the medieval iron-worker had in fact singled out most of the deposits low in phosphorus and rich in manganese. In an age when land-transport was difficult and water-transport, even when conveniently placed, was burdened with tolls and exactions, a local supply of fuel was essential. The charcoal-burner could cut over and burn vast areas of woodland, and very rarely, at least before the eighteenth century, was he careful to plant new seedlings. Lack of fuel, rather than lack of ore, was the most common reason for the closing down of iron works.

Italy, once important for its iron works, declined in importance as its thin forest cover was depleted. In the south of France the iron industry was only of slight importance because there was no assured supply of the only fuel that the early iron-workers could use. On the other hand, the hills to the west and east of the Rhine, in central Germany and in Austria, were well wooded. There seemed an abundant supply of fuel for an unlimited future. And man, im-provident as always, took no pains to conserve what he did not use and to replant where he had cut.

Wars took their toll of iron works. The expansion of the industry in the sixteenth century was cut short by military campaigns in the seventeenth. Germany suffered the most, and it seems that few works survived the Thirty Years' War. At least, scarcely any of the iron works that rose to fame and importance in the eighteenth century can be traced farther back than the generation that followed the Peace of Westphalia.

The expansion of the industry that was characteristic of the period of the Renaissance was revived and continued in the later seventeenth and eighteenth centuries. This growth was characterized by a further concentration in the few areas that were geographically well favoured. The advantages that came from abundant ore, and of fuel that was at least adequate for immediate needs, were matched by skills in iron-working that grew as the industry prospered. Developing regions attracted artisans from less prosperous areas. They also supplied craftsmen who took their skills to other regions. The Swedish iron industry in the seventeenth century owed much to the Flemish family of de Geer, who brought in the 'Walloon' pattern of refinery. In this way the iron industry of Western Europe was spreading outwards: to Scotland, Sweden, Finland and Russia. But its heart and centre continued to be in the belt of country that lies within a hundred miles of the river Rhine.

Coke-smelting and the Geographical Pattern: Hitherto the dominant factors in the rise and the decline of iron works had been the discovery and exhaustion of resources. In the late eighteenth century new technical processes began to affect the pattern.[2] First the use of mineral fuel instead of charcoal; then developments in steel-making that limited the varieties of ore that could be used; and, lastly, the practice of heat economy, and the growing scale of the undertakings themselves, combined to change the geographical pattern from that of the eighteenth century to that (Fig. 6) of the twentieth.

In the closing years of the eighteenth century mineral fuel began to be used on the continent of Europe in the smelting and refining of iron. Coke-smelted iron continued to have certain disadvantages and the prejudice against it continued in some parts of Europe far into the nineteenth century. The puddling process was under no such disadvantage. Though discovered in England more than seventy years after Abraham Darby's successful experiment with coke, it was practised on the continent several years before the first successful use of coke in the blast-furnace.

Both processes were extravagant of coal, a commodity very much more highly localized than charcoal. Charcoal was normally carried on pack-animals quite considerable distances, but the transport of coal was awkward unless the rivers could be used. The result was that the puddling industry was attracted to the coal-fields. As early as 1785 there was a puddling-furnace at Le Creusot, in central France, where coal was available. But the process did not spread until after 1815. Then Cockerill introduced it at Liège, on the Belgian coal-field. It spread quickly in France where, by 1827, there were no fewer than 149 puddling-furnaces.[3] In Germany it was adopted at Aachen, along the Rhine where the works could be supplied with coal by barge from the Ruhr coal-field, in Upper Silesia and then on the Ruhr coal-field itself.

The use of mineral fuel in the blast-furnace came later. Despite numerous attempts to smelt with coke in continental Europe, the only success that was achieved was in Upper Silesia. And there it was British technicians, brought across by the agents of Frederick

D

II of Prussia, who were responsible for it. In France coke was not introduced successfully until the 1820's, and its use did not become general until the second half of the century. In the Ruhr area the first successful use of coke fuel was in 1849, at the Friedrich-Wilhelm works in Mülheim.

The use of coke in the blast-furnace, like that of coal in the puddling-furnace, had the effect of attracting all branches of the industry towards the coal-fields. These early processes were extravagant of fuel, and to produce a ton of wrought-iron required far more coal than iron-ore. The ore moved to the coal. It is true that several coal-fields actually yielded significant quantities of blackband ore, notably the Ruhr coal-field itself. But such sources were supplemented with ores imported from northern Spain, Elba and Corsica, from the Siegerland and from central Sweden and northern England.

By the middle of the nineteenth century most of the older centres of production were in decline, and the industrial map was beginning to assume its present shape.[4] In France the industrial centres of Saint Étienne and Le Creusot were relatively as important as at any time in their history. Works had been established near the Alais coal-field in southern France, and on the coal-field of Nord and Pas-de-Calais. Savoy and the eastern Pyrenees continued their small-scale industry, but the iron industry as a whole had moved to the coal-fields.

In Belgium, the new industry established in the 1820's near Liège had spread westwards through the Belgian coal-field. The earlier works in the Ardennes were gradually abandoned[5] when Liège, Charleroi and Maubeuge developed as smelting and refining centres. In Germany the Siegerland retained its importance, but other centres of the older industry declined as more and more of the German industry was established on the two coal-fields of the Ruhr and Upper Silesia.

In Sweden the older style of industry continued with little interruption. Here, as in Finland, there was an abundance of timber, and charcoal remained the only fuel. In the Austrian Empire an older pattern of industry continued in the Alps of Styria and Carinthia, but in the Austrian provinces of Moravia and Bohemia, now part of Czechoslovakia, coal came to be used in the works at Vítkovice and near Kladno early in the nineteenth century.

Technology and the Geographical Pattern: Through the century the blast-furnaces were becoming larger and their management more skilled. The ratio of coal to pig-iron continued to decrease so that the economies of location near a coal-field became less conspicuous. The appetite of the puddling-furnace remained voracious, but in 1856 Bessemer introduced his converter and a few years later Siemens and Martin perfected the regenerative open-hearth furnace. These processes not only made steel directly from iron, but they did so with incomparably less fuel per ton of metal than the puddling-furnace had used. A further step was taken towards dissociating the steel industry from the coal-field.

It was, however, a technical problem which at this time started the movement away from the coal-field. The new steel-making processes could tolerate only a very minute proportion of phosphorus, and most of the European ores were moderately to highly phosphoric. The search began for ores exceptionally low in phosphorus, the so-called Bessemer ores, and they were found in the Basque province of Spain, in Elba, North Africa and central Sweden. As they were brought by sea to the consumer countries, the practice grew of erecting smelting works on the coast at their points of entry. This was particularly marked in France, a country almost devoid of Bessemer ore.

The results of this technical change had not worked themselves out when Sidney Gilchrist Thomas, with his use of a basic lining for the converter, found a means of using the new technological processes for pig-iron made from phosphoric ore. No longer was there a premium on imported, low-phosphorus ores, though Bessemer iron did not at once lose its importance. Eyes were turned rather to the immense deposits, in eastern France, in Luxembourg and in northern and western Germany, of low-grade, highly phosphoric ores. In France these ores had been known by the deprecatory name of *minette*. They had been used for generations, but nothing good had ever come of them. There was now a rush to use these ores. But their iron content was small and transport expensive. It was economic to bring fuel to the ore rather than the reverse. Within a few years new blast-furnace and steel works had been established in French and German Lorraine, in Luxembourg and near Peine and Salzgitter in the North German Plain. Ore from Lorraine and Luxembourg was transported to furnaces in Belgium, and some was taken to the Ruhr and the Saar, but the tendency grew to melt it on the ore-field.

The most significant change in the geography of the iron and steel industry in the second half of the nineteenth century was the rise of iron-smelting and steel-making in Lorraine and Luxembourg and, on a smaller scale, in Lower Saxony. These new plants used the basic converter, the so-called Thomas process, for making steel. The expansion of Thomas steel production was rapid. By the end of the century the iron-smelting industry was dominated by the furnaces erected on or very close to the fields of low-grade phosphoric ore. The economies of heat conservation dictated that much of the steel production should be carried on in the same area.

The Present Geographical Pattern:[6] We must guard against assuming that the geographicial pattern changed radically with each revolution in technology. The iron and steel industry adapts itself only very slowly to changing geographical conditions. Its plant and equipment is, as a general rule, not movable. It represents a huge capital investment which must be used if possible until it has been amortized. The expected length of life of a modern works is at least twenty-five years, and one can expect that, as technical modifications are made to it during these years, its life span must work out at a considerably longer period. In consequence the geographical pattern that emerged towards the end of the nineteenth century, and which has survived with only minor modifications, is a kind of palimpsest resulting from the superimposition of the several patterns dictated by changes in technology and resource base.

Individual plants and groups of plants located in accordance with the demands of the charcoal-iron industry survive in the Siegerland, though the local resources on which they first depended have entirely vanished. In the Ruhr, Belgium and northern France, as well as at Saint Étienne, Decazeville, Le Creusot and Aachen are works which owed their origin here to their excessive dependence on coal fuel. On the French coast are still a few works established to smelt imported Bessemer ore, and in Lorraine, Luxembourg and Lower Saxony are works which illustrate the movement, under the impact of the Thomas process, towards the easily worked but low-grade phosphoric ores.

In the last half-century new influences have been brought to bear

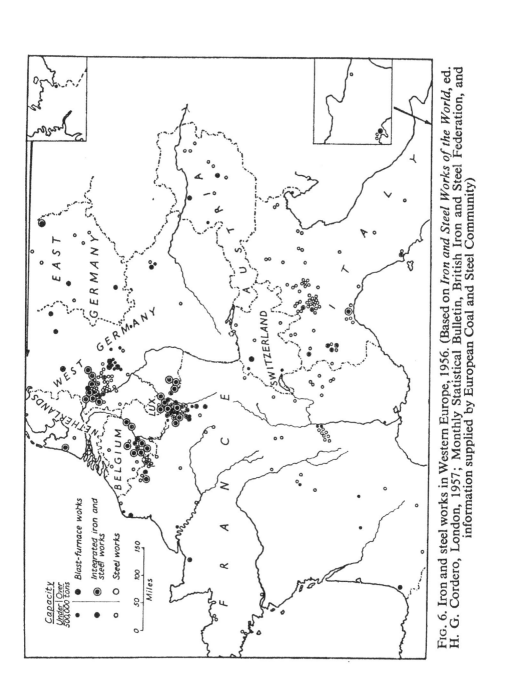

FIG. 6. Iron and steel works in Western Europe, 1956. (Based on *Iron and Steel Works of the World*, ed. H. G. Cordero, London, 1957; Monthly Statistical Bulletin, British Iron and Steel Federation, and information supplied by European Coal and Steel Community)

on the structure of the industry, modifying in detail its distribution in Western Europe. The open-hearth has gained a much wider acceptance, and in some industrial areas has completely displaced the converter. This arises in part from the fact that the steel which it produces is of a higher quality than Thomas steel, but also, in part, from the capacity of the open-hearth to consume scrap.

The steel-rolling industry and all branches of the steel-finishing industries produce large quantities of scrap. To this is added a growing stream of old scrap from dismantled plant and equipment. In this latter respect the industry could be expected to feel in the early years of the present century the consequences of the big expansion of steel production in the last quarter of the nineteenth century. The consumption of scrap in the open-hearth furnaces, and later in the electric furnaces, diminished relatively the need for pig-iron. Its effect was to give a certain cost advantage to steel works situated near those industrial centres, such as the Ruhr, Liège and northern France, where large quantities of scrap were being made available. Although no large steel works was able to operate without consuming pig-iron from blast-furnaces, a number of small works, not only in the older-established centres of France and Germany, but also in Italy, Switzerland and elsewhere, succeeded in maintaining themselves mainly on scrap, supplemented with pig-iron purchased in the open market.

A second trend, characteristic of the present century, reinforced further that which has just been discussed. Demand grew for quality steels, made in either the open-hearth or the electric furnace from pig-iron low in phosphorus and sulphur. Again, works in some of the longer-established centres—the Aachen district, Le Creusot and central Sweden—benefited. The use of electric power in the refining of quality steel had, as we have already seen in Chapter 1, immense advantages, but was as a general rule practicable only in areas where such power was abundant and cheap. The refining of electric steel has therefore gained ground in the French Alps, in Switzerland, northern Italy, Norway and Sweden. It remains relatively unimportant in regions where electric power is expensive—and these include most of the coastal sites and the industrial developments on the ore-fields. It has, however, made great headway in Germany in recent years.

Although Western Europe is relatively rich in iron-ores, most of these are low-grade and suitable only for making a high-phosphorus pig-iron. Such iron is suited neither to the open-hearth nor in the past could it be used for the manufacture of the higher qualities of steel. So, despite the very considerable reserves in eastern France and North Germany, the import of iron-ore from outside the West European area has increased, and now amounts to over a third of the consumption of Western Europe.[7] The most important source has been northern Sweden, supplemented by northern Spain, North Africa, Newfoundland and, in more recent years, sources even more remote in Africa, India and South America. Recently, however, the use of oxygen in the converter process has greatly raised the quality of the steel. This will in turn give increased importance to the types of ore most commonly found in Western Europe.

In consequence a fresh emphasis is being placed upon works located either on the coast or within reach by water-borne transport of the coast. Most of the works in the Ruhr belong to this group. They have good navigable waterways both to the port of Rotterdam by way of the Rhine and to Emden by the Dortmund-Ems Canal. New integrated works have been built at the North German ports. The Dutch iron and steel works at IJmuiden is accessible to sea-going ore-carriers; the French are building a new coastal works, near Dunkirk, and the newly rebuilt Italian industry is mainly coastal in its location.

A further influence on the location of the European iron and steel industry during the present century has been the movement towards national self-sufficiency. Steel was of too vital an importance for any country to be willingly dependent wholly on others for its supply. In several of the smaller countries its development was an aspect of the new nationalism. This is not to deny that in some of these countries there existed both the resources for the industry and a market for its products. It remains true, however, that when all allowances are made for specialization on a particular range of products, the scale of the resulting enterprise is too small to be economic.

The Italian and the Dutch steel works are among the larger of these autarchically developed industries, and to that extent more viable economically. The Italian has, nevertheless, been very highly protected, and the absorption of the Dutch into the Benelux union was not without its problems in the field of pricing. Spain, Portugal,

Greece, Denmark and Finland have developed and are, in some instances, in process of extending their iron and steel industries, despite the fact that the advantages to be derived from them certainly
do not lie in the sphere of low costs. The iron and steel industry of
Turkey, though it does not lie within the scope of this chapter, may
be cited as an example of an industry, uneconomic by Western standards, established primarily to induce a false sense of national well-
being.

It does not lie within the scope of this book to examine the
emergence of the large iron and steel concerns of Western Europe.
They began to form when, in the second half of the nineteenth
century, the manufacturers of iron and steel, in competition with one
another, began to acquire possession of their own reserves of ore and
fuel and then to reach forward into the finishing industries in order
to gain control of the market for their products. The formation of
such vertically integrated units had made great progress by the beginning of the twentieth century. Combinations then took place
between them. The iron and steel industry of the Ruhr, together
with a substantial part of its coal-mining industry, came in this way
to be controlled by only six large concerns. Similar concentrations
took place in France, Belgium, and Luxembourg, where a few concerns, notably the de Wendel interests, Schneider Frères, Cockerill,
Arbed and Ougrée-Marihaye achieved considerable proportions.

It is easy to see in such concentrations of economic power the
most sinister motives. But the concerns were developed in the
interests of economy and of low-cost production, not for the purposes of war and immense profits for the few. This is not the place
to pursue the argument that the great steel concerns were not built
up at the expense of peace and democracy, but were, on the con-
trary, a means of achieving economies in production and of smooth-
ing out demand both for raw materials and semi-finished goods.
We can, however, legitimately touch on their influence on the
geography of production.

The interests of the concerns of all the major West European
producing countries—Germany, Belgium, Luxembourg and France
—cut across national boundaries. To this extent these concerns had
a wider and a freer choice in the location of plant than might otherwise have been the case. Belgian and French firms co-operated in

Luxembourg industries, and Luxembourg firms in Germany and the Saar; de Wendel operated in both French and German Lorraine and the Ruhr concerns likewise in both Luxembourg and Lorraine and even in Normandy. Inevitably a kind of division of function occurred within each of these far-ranging concerns: smelting in Lorraine and Luxembourg and the export of pig-iron to steel-mills in the Ruhr, central Belgium or other parts of France. The pattern of movement of ore, fuel, pig-iron and steel within Western Europe could be varied within the limits set by the capacities of plant. In this way inefficient or uneconomic plant could be closed temporarily or permanently and greater economies achieved than would otherwise have been the case.

In 1926 the International Steel Cartel was formed.[8] This was an association between the leading producers with the object of controlling production and prices in relation to demand. It imposed national production quotas, but did nothing directly to make production cheaper or to locate production in the most favourable places. Its effect was rather to stereotype an existing geographical pattern.

The European Coal and Steel Community: The trends outlined in the previous pages have been continued since the Second World War. The German industry, fragmented by the action of the Allied Powers, has fused again to form concerns similar to those existing before the war. But the International Steel Cartel has not been revived. The European Coal and Steel Community, which came into being with the ratification of the treaty between its six member States in 1952, is in two important respects the opposite of the older cartel. It was created, in the first place, by the action of the governments, not of the industrial firms, in its member countries. Secondly it does not operate on a quota basis, and thus does not tend to stereotype an existing pattern. Its purpose is not to maintain prices by restricting output, but rather to force costs down by creating conditions of free competition between the industries of all its member nations. By the removal of all artificial causes of friction in the movement of ore, metallurgical fuel, scrap, pig-iron and crude and rolled steel, the Community seeks to ensure that the industry grows where essentially geographical conditions are most suitable for it to develop efficiently and economically.

In addition to its primary purpose of removing the artificial restrictions on trade in steel and steel-making materials, the Community watches the industry as much from the point of view of the consumer as from that of the producer. If steel-making scrap is scarce, then it attempts to assure supplies and keep prices down to reasonable levels. If new smelting, steel-making or rolling capacity is needed, then the Community does its utmost to secure that the additional works are so located as to achieve all advantages possible from geographical position. A study of the annual reports of the High Authority of the European Coal and Steel Community shows how an international industry is being controlled and directed by an international authority in the interest of the cheapest production of steel goods. From the geographical point of view the most interesting aspect of the operation of the Common Market for Coal and Steel is that it will assist in locating new plant, not in the best situations for a particular country, but in the best from the point of view of the six countries of the Community as a whole.

European Iron and Steel Output: In the following pages, a brief survey is undertaken of the iron and steel industries of each of the countries of Western Europe. The table gives a picture of the relative importance of these countries and shows also the extent of their expansion since the pre-war period:

	Average 1936–8		*1965*	
	Pig-iron	*Steel*	*Pig-iron*	*Steel*
		(thousands of tons)		
France.. ..	6,682	6,950	16,020	19,604
Belgium ..	3,130	3,121	8,366	9,169
Luxembourg ..	2,017	1,976	4,145	4,585
Netherlands ..	284	44	2,364	3,138
West Germany ..	13,823*	16,181*	26,990	36,821
Italy	800	2,145	5,625	12,660
Spain	265	371	2,394	3,460
Switzerland ..	—	12	27	345
Austria ..	396	579	2,205	3,221
Denmark ..	—	22	78	412
Norway ..	35	65	1,080	686
Sweden ..	635	1,033	2,446	4,725
Finland ..	16	55	984	363

(*Quarterly Bulletin of Steel Statistics.* and *Minerals Yearbook*, U.S. Bureau of Mines.)

* Boundaries as of 1936.

Since the end of the Second World War the trend in European steel production has been steadily upward. The works were rebuilt or reconditioned, following the destruction and neglect of the war years, and in all countries there were ambitious plans for further extension of the industry's capacity. Trade between the countries of Western Europe is large, but the area as a whole was always heavily dependent on exports to non-European markets. The less-developed countries were also expanding their steel capacity, and concern was expressed regarding the future of the export market for steel. A report, published in 1949 by the Economic Commission for Europe on these trends,[9] pointed out that the extra-European market would be able to take less than half of Western Europe's export surplus. The possibly serious consequences of an over-expansion of the European industry were anticipated by the Korean crisis and a renewed expansion of armaments programmes. Furthermore, it seems likely that the authors of this report had underrated the volume of overseas consumption and the effect of heavy investment in the underdeveloped countries. Apart from a recession, largely confined to the United States, in 1954, the steel market has continued its upward trend, and the confidence of the steel-producing countries in the future of the market is demonstrated by their plans for expanding their steel capacity.

France: The problems of the French iron and steel industry derive in part from its long history, in part from the shortages of certain raw materials.[10] Its organization and equipment were not satisfactory before the Second World War. During the war itself plant suffered from destruction and neglect. In 1946 the first Monnet Plan charted a programme for an extensive retooling, accompanied by a modest expansion of the industry. This was followed by a second plan, which, in harmony with the expanding market, called for considerable additions to French iron- and steel-making capacity.

The geographical pattern of iron- and steel-making in France has undergone radical changes during the past century and a half. But the older patterns have in no instance been completely obliterated. Always some works survive from an earlier age. One of the problems of the French industry was the large number of very small and inefficient units. The modernization plans have focused their interests on the large and well-placed plants. As the industry continues to expand, the older centres become relatively of smaller importance and some of them may be expected in time to close.

France has, as we have already seen, very large reserves of low-grade, phosphoric ore, but very little ore of higher quality. Her coal resources, located mainly in the north and in Lorraine, are not in general suited to the making of metallurgical coke, and France has had to rely heavily on imported fuel, chiefly from Germany. These factors have greatly influenced the location of plant in recent years. Works on the ore-field of Lorraine now produce about 78 per cent of France's pig-iron and about 66 per cent of the steel. The current expansion of capacities in Lorraine will increase these percentages.

The most extensive coal resources in France are also in Lorraine, where they form an extension of the Saar coal-field. The coal has, however, never been considered of good coking quality, but recent experiments in blending the Lorraine coal with fuel from other sources have met with a large measure of success.[11] It seems, how-ever, that some dependence must continue to be placed on German coking coal, and Lorraine firms have recently acquired a group of mines in the Ruhr.

The Lorraine plants are mostly situated just below the limestone scarp in which the ore occurs. They could not easily be nearer their source of ore, but are at a considerable distance from the ports and from much of the domestic market. The canalization of the river Moselle, agreed on by France and Germany, will undoubtedly facilitate the shipment of products, though it is aimed at lowering the cost of importing German coal.[12]

The north of France is second in importance to Lorraine, with 14 per cent of the pig-iron and 22 per cent of the steel production. It is not an important smelting region, and some pig-iron is brought in from Lorraine for steel-making here. The works lie mostly near Valenciennes, but a new works is planned on a coastal site near Dunkirk.

The centre and south of France embrace the oldest centres of the French industry. Much of the plant is old, and only a few years ago one might have said that there was little future for the iron and steel industry in this part of France. Indeed, there has been in recent years a decline in iron-smelting and also of steel-making by the open-hearth process. The works are old and small, and some have closed recently. On the other hand, the expansion of electric steel has been such that the overall steel production of the area has increased. No less than twenty-three works make electric steel, and a considerable expansion of production is foreseen in the near future.

Luxembourg: The industry of Luxembourg resembles that of Lorraine, in being based on the local reserves of *minette*. It differs, however, in its complete dependence on its neighbours—mainly Germany—for the supply of fuel. Luxembourg contains only a small part of the *minette* deposits, and doubts have been expressed regarding the adequacy of ore reserves. These are, however, adequate for the life of the present works, though it is doubtful whether any great expansion would be wise without assuring a supply of ore from the French sector of the ore-field. Luxembourg already uses French ore in quantity to blend with her own much poorer ores, and future supplies are assured through the operation of the Common Market.

The domestic market in the Grand Duchy can absorb only a small fraction of the iron and steel production, and Luxembourg is dependent on the export market.

Belgium: The Belgian iron and steel industry grew vigorously during the middle years of the nineteenth century, but more recently its rate of growth has slowed. Liège and Charleroi have been the chief centres from the start of the modern industry. Coal was obtained from the coal-field of the Meuse and Sambre valleys, and iron-ore formerly from the Ardennes. Belgian coal-mines are now expensive and inefficient, and ore has to be imported, partly from Lorraine and Luxembourg, partly from overseas.

Netherlands: The single integrated iron and steel works in the Netherlands was established in 1924 at Ijmuiden, to the west of Amsterdam. It was not however completed before the Second World War began. Ore is imported, chiefly from Sweden and North Africa, and coking coal is brought by barge from the Limburg field in the southern Netherlands. The works are unusually well served by water transport. The high rate of scrap production in the Netherlands had encouraged the use of open-hearth furnaces. In addition to the major works at Ijmuiden there are also open-hearth and electric steel works at Utrecht.

Germany: Germany has fully regained her pre-war position as the biggest iron and steel producer in Western Europe. She has advantages, in addition to the immense industry of her people, that are denied to most of her neighbours. Her reserves of iron-ore are not very large, and, like those of France, are almost restricted to low-grade, phosphoric deposits. But the Ruhr coal-field is not only the richest in Europe in terms of total resources, but has also the highest proportion of coal of coking quality.[13] On it depend in some degree the smelting industries of Luxembourg, France, Austria and Italy. The Ruhr industrial area, furthermore, is well placed for the trans-port of its raw materials and products. Apart from a well-developed railway net the Ruhr area has at its western end the navigable Rhine. Stretching from west to east through the region are the Herne and Lippe canals. At the eastern end these merge with the Dortmund-Ems Canal which provides for water-borne transport to the North Sea port of Emden.

The smelting and steel-making works are clustered at the eastern and western ends of the Ruhr area. Most are near the Rhine port of Duisburg-Ruhrort, but three of the largest lie near Dortmund. Except in the latter vicinity, they are placed close to either river or canal, so that ore is unloaded directly from barge to stockpile. The ore used comes mainly from northern Sweden by sea and is trans-shipped to barge at Rotterdam or Emden. Limestone is obtained from the hills to the south of the Ruhr, and the fuel has at most to be brought a mile or two from the mine to the plant.

Between the eastern and western extremities of the Ruhr area, steel-making and steel-rolling and fabricating are more strongly developed than the smelting branch. Essen is no longer a centre of iron-smelting, and concentrates on making quality steels at the Krupp works. Bochum specializes in steel-castings and other plant in this central area in high-grade steels.

The Ruhr has about 80 per cent of Germany's pig-iron and steel capacity. There are steel and rolling-mills to the south-west of the Ruhr, at Krefeld and Düsseldorf, but the iron industry, formerly widespread in central Germany, is found now only in the Siegen district and at Amberg in Upper Bavaria.

Here the smelting and refining industry arose in the later Middle Ages, using the high-quality ores of Müsen and the abundant fuel from the forests. The local ore supply is no longer important, and all materials have to be brought into the area, but a small iron-smelting and steel-making industry survives. Some of the blast-furnaces are small and old and produce only foundry-iron, but, for the rest, the Siegen industry is vigorous and, despite severe geographical disadvantages, expanding.

The most significant development in the German industry in recent years has been the creation of the Watenstedt works in the Lower Saxon plain. Here are considerable reserves of *minette*-type ores, and during the nineteenth century several works were established to smelt them, notably at Osnabrück, Ilsede and Peine. During the Hitler period a large integrated works was built as part of the Hermann Goering concern. Despite war damage and dismantling, these works have been extended and now constitute one of the largest in Germany.

The possession of large reserves of excellent coking coal allows the German industry to achieve greater economies than other West European countries. Only the shortage of good-quality ore gives cause for alarm. But the Germans are interesting themselves in the mining developments of Labrador, South America and elsewhere. In peacetime at least their ore supply should be secure.

Italy: Italy is, in terms of resources for the iron and steel production, one of the weakest countries of the European Community.[14] The total iron-ore reserves are small, with the most valuable on the island of Elba. Italy does not hope to satisfy more than a half of her needs from domestic sources. The scanty coal deposits do not produce any fuel of metallurgical quality. All coking coal has to be imported, though the relatively large consumption of electric power and natural gas does to some extent diminish the need for coal.

Before the war the industry was small and, partly through mismanagement by the Italian government, by no means efficient or profitable. It has been rescued from this plight since 1945 and its plant has been greatly extended. It remained, however, highly protected, though the customs were gradually reduced and, by the terms of the Community agreement, were abolished in 1958. The present output of pig-iron and steel is rising sharply, despite the danger of excessive dependence on imported materials. In particular, Italy imports very large quantities of metal scrap, chiefly from the United States.

The integrated works lie, with one exception, on the coast at Cornigliano (Genoa), Piombino (Tuscany), Bagnoli (Naples) and Trieste. The exception is near Aosta in the Alps of Piedmont. Steel and rolling works are very widespread, though most numerous near Genoa, Turin and Milan. In addition to the major works, numerous small plants exist. Many use electric steel-furnaces and consume very considerable quantities of metal scrap, some of it made available by the engineering industries of northern Italy, the rest imported.

Switzerland and Austria: The Alpine region contains numerous small deposits of iron-ore, is rich in electric power, but almost completely lacking in mineral fuel.[15] Furthermore, the strong relief makes the development of transport facilities difficult and waterborne transport impossible.

Switzerland's iron industry is interesting because of the extensive use that is made of electrically fired, low-shaft furnaces. Nevertheless, the annual production of pig-iron is only about 40,000 tons. Steel production is, of course, very much greater. The scrap-metal supply, even from Switzerland's extensive metal-using industries, is insufficient, and has to be supplemented by imports of scrap and of pig-iron. Steel production is entirely by electric furnace and is, in consequence, generally of a high quality.

Austria is better endowed than Switzerland, and contains high-quality ores and a long tradition of iron-working. This was centred in the provinces of Styria and Carinthia, where were small but rich ore deposits.[16] The Erzberg of Styria has been mined continuously at least since the Middle Ages, and, with its giant steps cut into the steep mountain side, forms one of the most spectacular iron 'mines' in the world. Most of the earlier iron works are closed, and the region is dotted with the ruins of long-abandoned furnaces. There remain two modern, integrated iron and steel works, at Donawitz, a few miles from the Erzberg deposits, and at Linz on the river Danube.[17] There are numerous steel- and rolling-mills along the valleys of the Mur and Mürz, which use pig-iron and scrap. Fuel is imported from Germany by rail, but ore costs are low, and the Austrian steel industry is now being modernized and extended.

Spain and Portugal: Like Austria, Spain has a long tradition of iron-working, and some of the nomenclature of the older industrial processes witnesses to a Spanish origin.[18] But Spain is not well endowed for the modern practice of steel-making. The rich ores of Vizcaya, which were in such great demand in the nineteenth century for the making of Bessemer iron, are nearing exhaustion. Spain has numerous other sources of ore, but these are either poor in quality or difficult of access. It appears that the future industry will be based on the low-grade and highly siliceous ores of Asturias and Leon. The most extensive Spanish coal-field also lies in Asturias, inland from the city of Gijon. The coal is poor in quality and little of it is of coking quality, and it has to be supplemented by imports of coke and coking coal.

The blast-furnace and steel works lie mostly on the north coast, at Bilbao, Santander and on the Asturias coal-field. There is a single works on the Mediterranean coast, near Valencia. Eight integrated works produced in 1956 944,000 tons of pig-iron and 1,243,000 tons of steel, a fact which suggests strongly that the plants were neither modern nor efficient. The most ambitious industrial plan of the Franco regime has been to build a modern works at Aviles, on the coast near Gijon. The first furnace of this works was blown in late in 1957, and the Aviles works have already almost doubled the steel capacity of Spain.

Portugal is less well-endowed even than Spain. At present there is only a very small steel production, based on imported pig-iron and scrap, but plans are on foot for the extension of the industry, perhaps by using one of the recently devised sponge-iron processes.

Scandinavia: The ancient rocks of Northern Europe are richly furnished with iron-ore, but their great age precludes the possibility of finding coal. On the other hand their forests still provide charcoal for the iron industry, and their immense reserves in hydro-electric power open up the wide field of electric metallurgy. This interesting combination of the very old with the very new is illustrated by the following table of pig-iron production in 1962:

	Charcoal-Iron	*Coke-smelted Iron*	*Electrically smelted Iron*	*Total*
		(thousands of tons)		
Sweden	26	1,594	142	1,812
Norway	—	—	399	399

The change from the old to the new has been very recent.[19] In 1900 Sweden had 166 blast-furnaces, all of them using charcoal, and producing about half a million tons of metal. Output has more than doubled since then, but at least three-quarters of the present total is now from coke-fired furnaces. Charcoal-iron, smelted from the pure ores of central Sweden, is used for making steel of the highest quality. The ratio of scrap to pig-iron in the steel-making processes is high, and almost a half of the steel is made in some form of electric furnace.

The most interesting feature of the Swedish industry is the complete dichotomy between her large reserves of high-grade ore and her small production of quality iron and steel. The ore from the vast deposits of northern Sweden (*see* page 40) is exported. The domestic industry is based on the very much smaller ore deposits of central Sweden and on large quantities of metal scrap. The domestically used ore is from the Bergslagen district, to the west and north-west of Stockholm, and here, in the ancient centres of the Swedish industry, are the numerous smelting and steel-making plants.[20]

In Norway, the small industry is heavily dependent on electric power. The pig-iron is produced exclusively in the electrically fired blast-furnace, and most of the steel is also made in the electric furnace. There is, as might be expected, a high ratio of scrap to pig-iron.

The industries of Denmark and Finland, both very small, make some use of the orthodox blast-furnace, but the following table, showing materials consumed in 1961, shows how dependent is the industry on scrap used in the open-hearth:

	Pig iron used	Scrap used	Spiegeleisen and ferromanganese	Steel made
Denmark	109·2	291·5	1·2	367
Finland	57·3	301·7	2·9	331

(*thousands of tons*)

Turkey: The Republic of Turkey is so closely associated politically with the countries of Western Europe, that it seems more appropriate to consider the Turkish iron and steel industry here, rather than with the underdeveloped countries of Asia.[21] A single integrated works was built in the 1930's at Karabük, near the north coast of Asia Minor. The wisdom of this step has often been questioned. There are nearby reserves of coking coal, but the iron ore of Divrik, in eastern Turkey, is small in amount, distant from the works, and not of a high grade. The works themselves are small, and their output in 1963 was only 210,000 tons of pig-iron and 331,000 of steel.

[1] John V. Nef, 'Mining and Metallurgy in Medieval Civilization', *The Cambridge Economic History of Europe*, Vol. II, Cambridge, 1952, 430–92.

[2] N. J. G. Pounds and W. N. Parker, *Coal and Steel in Western Europe*, London and Bloomington, Ind., 1954. Much of the following pages summarizes this book. Sources are quoted in this work.

[3] *Annales des Mines*, 2nd Series, II, 1827, 401–70.

[4] A valuable article on the geographical change is Marcel Bulard, 'L'Industrie du Fer dans la Haute Marne', *Annales de Geographie*, XIII, 1904, 223–42, 310–21.

[5] René Evrard, *Forges Anciennes*, Liège, 1956.

[6] See *The Iron and Steel Industry in Europe*, O.E.E.C., Paris, 1956; Maurice Fontaine, *L'Industrie Sidérurgique dans le Mande*, Paris, 1950; *Gemeinfassliche Darstellung der Eisenindustrie*, Düsseldorf.

[7] *Fifth General Report of the Activities of the Community*, E.C.S.C., 1957 90–91.

[8] Erwin Hexner, *The International Steel Cartel*, Chapel Hill, N.C., 1943.

[9] *European Steel Trends in the Setting of the World Market*, Economic Commission for Europe, U.N., Geneva, 1949.

[10] J. Chardonnet, *La Sidérurgie Française*, Paris, 1954. For the earlier period see *La Sidérurgie française 1864–1914*, Comité des Forges, Paris, 1914, and references cited in N. J. G. Pounds and W. N. Parker, *op. cit. See also* J. E. Martin, 'Location Factors in the Lorraine Iron and Steel Industry', *Institute of British Geographers, Transactions and Papers*, 1957, 192–212; *ibid*, 'Recent Trends in the Lorraine Iron and Steel Industry', *Geography*, XLIII, 1958, 191–199.

[11] *Recent Advances in Steel Technology and Market Development*, 1954, U.N., Geneva, 1955, 6–10.

[12] See N. J. G. Pounds, 'Lorraine and the Ruhr', *Economic Geography*, XXXIII, 1954, 149–62.

[13] N. J. G. Pounds, *The Ruhr*, London and Bloomington, Ind., 1952.

[14] John Cairncross, 'The Future of Italy's Steel Industry', *Banca Naziona le del Lavoro, Quarterly Review*, Sept., 1957, 352–68; 'The Italian Iron and Steel Industry', *Monthly Statistical Bulletin*, B.I.S.F., April, 1952.

[15] 'The Swiss Iron and Steel Industry', *Monthly Statistical Bulletin*, B.I.S.F., April, 1953.

[16] The history of this industry is well documented in *Beitrage zur Geschichte des österreichischen Eisenwesens*, Vienna, 1931, *et seq.*

[17] 'Austria's Steel Investment Programme', *Monthly Statistical Bulletin*, B.I.S.F., August, 1951.

[18] 'The Spanish Iron and Steel Industry,' *Monthly Statistical Bulletin*, B.I.S.F., January, 1953.

[19] Gunnar Lowegren, *Swedish Iron and Steel: A Historical Survey*, Stenska Handelsbanken, Stockholm, 1948.

[20] A very useful map appears in *Statistical Year Book for 1954* of the B.I.S.F., Overseas section, London, 1956, 332–33. 'Sweden Enlarging Output of Quality Steel', *Times Review of Industry*, Sept., 1958.

[21] E. Tümertekin, 'The Iron and Steel Industry of Turkey', *Economic Geography*, XXXI, 1955, 179–184.

6

THE IRON AND STEEL INDUSTRY
OF THE UNITED STATES AND
CANADA

IN THE 1880's the output of the blast-furnaces of the United States became the biggest in the world, and before 1900 American steel production pulled ahead of that of its nearest rival, Germany. Since that date the iron and steel industries of the North American continent have expanded at an extraordinary pace. Production in Canada, begun more than half a century ago, has added its contribution to the growing American total. In 1957 the United States and Canada produced together 25 per cent of the world's pig-iron and nearly 30 per cent of its crude steel. Their closest rival, the Soviet Union, produced considerably less than half this total.

This powerful growth on the North American continent has been assisted by many factors, both geographical and social. The vast reserves of easily exploited ore; the high quality and the abundance of coking coal; the convenience of the Great Lakes system of water transport are all part of the rich natural endowment of the continent. The network of man-made communications, the market of between 100 million and 200 million people throughout this period of rapid growth, the freedom from restrictions on trade and movement, and the high level of consumption of most of the population have influenced deeply the course of this expansion.

Steel goods have come to enter deeply into the daily lives of the people. The cities are made up of skeletons of steel, draped with concrete and masonry. The bridges and the raised viaducts absorb huge quantities of steel; sheet metal enters the home in the shape of domestic appliances in a way that is unknown in other continents. There is a car, containing on average a ton and a half of steel, for every two and a half people. Not only is the demand great, but the turnover is also large. Most of these goods have a short life, and their cycle is completed when they pass through the scrap-dealer's yard back to the open-hearths of the steel-mill from which they came.

A large domestic market has encouraged production on the largest scale, and this in turn has permitted economies which have had the effect of extending yet more widely the market. To these advantages most Americans would add 'the continued maintenance of freedom to compete'[1] between existing firms in the industry and the encouragement given to experiment by the private enterprise system.

The Early Iron Industry in North America: There is no reason to suppose that the North American Indians had acquired the art of iron-working before the arrival of Europeans in the sixteenth century. But within a dozen years of the establishment of the first permanent English colony at Jamestown, iron-ore was being smelted. The earliest works were built on Falling Creek, a tributary of the James river.[2] They had, however, only a short life, and were soon afterwards destroyed in an Indian raid. It was many years before iron-working was begun again in Virginia.

In the meantime, however, a very successful furnace was built on the Saugus river, in Massachusetts.[3] It smelted the local bog-ores obtained from the meadows around Lynn. The iron industry spread slowly through the hills of New England, where there was charcoal in abundance and an at least adequate supply of easily worked ore.[4] In the hills of western New England larger deposits of the more valuable haematite were found. Foundries were established to cast cannon, as well as refineries to make the metal agricultural tools and equipment.

Towards the end of the seventeenth century iron-furnaces were built in New Jersey, where, as in Massachusetts, there was no lack of both ore and lumber to make charcoal.[5] From here iron-working spread northwards into the hills of New York State and westward into Pennsylvania. Pennsylvania, still today the most important State in the American steel industry, had its earliest furnace by 1692. The first permanent and successful iron works came a few years later, near Pottstown, on the Schuylkill river. Others followed in the next decades amid the rolling hills, with their rich forest and scattered beds of haematite, of eastern Pennsylvania. Amongst these works was the Valley Forge, in whose vicinity Washington and his continental army spent the bitter winter of 1777. A belt of country, stretching north-eastward from the Susquehanna river to the Delaware, and including the cities of Lancaster, Lebanon, Reading, Pottstown and Bethlehem, came, before the end of the eighteenth century, to be dotted with furnaces and refineries. Not only did they supply the military and agricultural needs of the colonists, but even shipped pig- and bar-iron to Europe, to the chagrin of the English iron-masters.

To the west of this, the earliest and for long the most important of American iron-producing regions, lay the higher ridges of the Appalachian mountains. These were traversed in this vicinity by the Delaware, Schuylkill, Susquehanna and Juniata rivers, each flowing in turn along the wide valleys between the ridges and across the ridges by narrow and steep-sided valleys.

In the latter half of the eighteenth century the iron industry was beginning to spread up these valleys. Bog-ore was found along the valleys and haematite in small deposits in the hills, and the supply of good hardwood for making charcoal seemed inexhaustible. Much of the iron produced amid these hills was floated down the rivers to the east-coast cities where it found a market.[6]

The Allegheny mountains here form the divide between the Atlantic and the Mississippi drainage. Iron-ore was found to the west of the watershed in the 1780's, and within a few years a bloom-ery had been built in the Monongahela valley. Fayette County, which contains Uniontown and Connellsville, was soon as busy as the Juniata or Schuylkill valleys. The industry spread northwards through the steep, wooded hills, drained by the Youghiogheny and Conemaugh rivers.

Once they had pressed beyond the divide, the iron-workers could no longer find an outlet for their goods in Philadelphia and New York. The high cost of overland travel prohibited that. The alternative was to allow their trade to follow the rivers which flowed to join the Ohio. These converged at the Forks of the Ohio, where the city of Pittsburgh now stands. A blast-furnace was built within the limits of the present city as early as 1792, but it worked for only a year or two. There seemed at this time no future for a smelting industry in Pittsburgh. But pig- and bar-iron from all the furnaces and forges west of the Allegheny mountains came through the city. Settlement was spreading through the Midwest, and the demand was growing for axes, saws, nails and for all the tools and equipment of the pioneer. Pittsburgh became an iron-using centre. A foundry was established about 1805, followed by several nail-making and tool-making factories and rolling-mills. Throughout the first half of the nineteenth century Pittsburgh continued to grow not as an iron-making but as an iron-using centre, drawing its materials from the numerous works that lay up in the valleys of the Allegheny Plateau.

While Pittsburgh was developing its fabricating industries, blast-furnaces and refining hearths were spreading north-westward through the Mahoning valley, where are now the steel works of Youngstown, to the shore of Lake Erie, and south-west, down the Ohio river, where a centre of the industry grew up at Ironton and Ashland, on the borders of Ohio and Kentucky. Charcoal from the woodlands and local ores, often bog-ores, provided the materials. But these works showed no great vitality. As the land was brought increasingly under cultivation the source of charcoal was cut off, and the supply of ore was exhausted within a short period of time. By the late-nineteenth century most of these furnaces had closed. Though other charcoal-iron works sprang up sporadically in Indiana, Illinois and other States of the Midwest, the region as a whole came to depend more and more on the Pittsburgh region for its supply of metal and metal goods.

In the eighteenth century an iron-smelting industry, based mainly on the bog-ores found in the coastal plain, spread southward from New Jersey and eastern Pennsylvania. It achieved some importance in Maryland and Virginia, but further south the works were small, and their methods and equipment were primitive. In general they satisfied only some of the needs of their own locality, and their owners did not engage in any long-distance trade in the products of their industry.

The Use of Coal and Anthracite: Charcoal remained the dominant and, in most areas, the only fuel used in iron-working until the middle of the nineteenth century. The abundance of charcoal resources had permitted and the lack of developed means of transport had dictated a widely scattered geographical pattern of iron-working. But by about 1850 the railway network was spreading rapidly over at least the eastern half of the country, and, locally at least, the sources of charcoal and ore were becoming exhausted. The speedier means of communication permitted the more rapid diffusion of technical knowledge, and new developments, sometimes of European origin, in blast-furnace design and steel-works operation spread to the centres of the American industry.

The result was the gradual abandonment of charcoal in favour of coal fuel; of the puddling and rolling of wrought-iron for the refining of steel, and the intensification of the industry at the few centres geographically most suited to it. This phenomenon is already familiar in Europe, but it happened more recently in America and, to that extent, can be followed more closely.

In 1839 a furnace was built at Pottsville, in eastern Pennsylvania, in order to use anthracite as fuel. A few years earlier, the hot blast, invented in Scotland by Nielson, had been introduced here. Woodland was by this time becoming scarce in eastern Pennsylvania, and the only available mineral fuel was the anthracite from the Scranton coal-field. As in Scotland, the higher temperature in the blast-furnace, produced by the use of the hot blast, permitted the sulphur to be dispelled more easily and crude coal to be used. The use of anthracite allowed the eastern Pennsylvania region to develop until it became for a short period the most important centre of the iron industry in the country.

The use of anthracite as blast-furnace fuel continued to increase until the end of the century, but after 1875 it was exceeded in importance by coke made from 'soft' or bituminous coal. The use of anthracite had established the focus of the iron industry in eastern Pennsylvania. The discovery of large reserves of high-quality coking coal shifted this focus westwards to the Ohio valley.

Early experiments in the use of coke for smelting seem to have been uniformly unsuccessful, and it was not until after 1850 that it began to be used at all widely. The expansion of coke-smelting is indissolubly connected with the opening up of the Connellsville field of western Pennsylvania. Here was a coal capable of yielding a hard, firm coke, almost wholly free of sulphur, and admirably suited to blast-furnace use. It was mined as early as the 1840's, and in the 1850's began to be used in iron works. It was not until about 1860, however, that its use spread to Pittsburgh.

At Pittsburgh, in the meantime, iron-smelting had been revived. For the first half of the nineteenth century the city had been a centre of foundry work and iron-rolling, but in 1859 a blast-furnace was built. Connellsville coke brought down the Youghiogheny river was substituted for the local fuel, and the furnace ran with great success. Within a few years many other furnaces were built, and within thirty years Allegheny County, the administrative unit within which the city of Pittsburgh lies, was smelting more iron than any State, except Pennsylvania itself, in the Union.

This rapid rise of Pittsburgh as an iron-smelting and steel-making centre was based, first and foremost, on the local reserves of good coking coal and only secondly on the iron-ores of the district. Indeed, the latter were nearing exhaustion when the industry was growing most rapidly, and the iron-masters had soon to turn to other sources. But means of transport and communication were good. Pittsburgh lay at the focus of navigable rivers which drained the hill country of western Pennsylvania, and the river Ohio offered an avenue for river transport south-westward to the Mississippi. A growing network of railways linked Pittsburgh with the older industrial centres of the east and with the growing markets of the West.

The discovery of Connellsville coal and with the simultaneous discovery of iron-ore along the shores of Lake Superior offered new possibilities. Several years elapsed before the iron-ore transport from Lake Superior was organized (*see* below, page 125). But with the beginning of the regular shipping service for ore, it became profitable to establish smelting works in and near Pittsburgh. Here, ore brought by rail from Lake Erie ports met the coke that came by barge down the rivers, and the pig-iron found a market in the rapidly growing puddling, steel-making and rolling works.

This growth in Pittsburgh itself owed much to the driving power and the organizing capacity of Andrew Carnegie. He became interested in the iron business in 1863, when he was twenty-eight. Within a few years several works were organized around himself until he was able to merge them into the Carnegie Steel Company.[7] He then absorbed the significant rivals in this area into his own company. He built blast-furnaces in Pittsburgh to supply his steel works; he absorbed the largest of the Connellsville coke companies to supply them with fuel and got control of a lion's share of the Lake Superior ore deposits to ensure a steady flow of ore. He owned or controlled steel-using firms to provide a secure market for the products of his steel-mills. When in 1901 Andrew Carnegie withdrew from the industry, his company became the nucleus of the United States Steel Corporation. In his later years he dominated the whole industrial structure of Pittsburgh and was the most powerful steel 'magnate' in the world.

Integrated blast-furnace and steel works were also growing up further down the Ohio river in the later years of the nineteenth century, deriving their coke from Connellsville and their ore from the shores of Lake Superior. Wheeling was the most important of these. Like Pittsburgh, it had grown up as an iron fabricating centre, and from the first it specialized in nails, thousands of tons of which were carried westward by boat and covered waggon to build cabins on the widening frontier. But smelting and then steel-making were introduced, and the rolling of sheet and rails was established.

A second industrial centre, peripheral to Pittsburgh itself, was in the Mahoning and Shanango valleys. These discharge southwards to the Ohio river, but were never navigable on the scale of the Monongahela and Youghiogheny. Connellsville coke might be a little more expensive here than in Pittsburgh, but ore from Lake Superior cost less. Industry grew up here at the same time as in Wheeling, and the city of Youngstown came to be dominated by the iron and steel industry.

By about 1900 an area within sixty miles of Pittsburgh, and including Youngstown and Wheeling, was producing some 40 per cent of the iron and steel produced in the United States.[8] The older centres of production in eastern Pennsylvania were, relatively at least, of diminishing importance. The small works scattered through the Midwest and the South were disappearing. Pittsburgh had the best and the cheapest fuel.[9] Ore came from distant Lake Superior, but the scale of mining and the organization of transport did much to reduce its cost. Pittsburgh enjoyed the most developed system of transport in the country, and its industry was directed by a group of aggressive industrialists and financiers of unparalleled ability. Lastly, this ascendency of the Pittsburgh district was perpetuated by the system of charging for its products that came to be known as 'Pittsburgh plus'. In 1900, the costs of production of iron and steel goods were lower in Pittsburgh than anywhere else in the country. Other producing centres might be nearer the market, but costs of production would be higher. The practice was adopted in 1900 of charging everywhere for each type of iron and steel product the current price in Pittsburgh *plus* the freight from Pittsburgh to the market wherever the producer and the market might be.[10] No great hardship was wrought on producers in other parts of the country as long as their prices remained high. But as soon as their costs of production fell below those of Pittsburgh, as they were soon to do in the Chicago area, the effect of 'Pittsburgh plus' was to weaken competition and give added power to Pittsburgh business. Until it was declared illegal in 1924, this pricing system discouraged the geographical spread of the industry and strengthened its concentration in the Pittsburgh area. It was replaced by a multiple basing-point system, which still gave some protection to the older centres, and lastly, in 1948, by a system of quoting prices f.o.b. at works.

The Lake Superior Iron-ore Deposits: Several references have already been made to the iron-ore deposits of the upper Great Lakes, and to their importance for American industry. They occur, as was described in Chapter 2, in a series of 'ranges' which lie roughly from south-west to north-east and are relatively long compared with their breadth. The first to be found was the Marquette Range of Upper Michigan, discovered in 1845 by a surveyor. The ore quickly attracted attention, and within a few years several companies had been formed and had staked their claims to iron-bearing land.[11]

At this date the transport of ore from the mines to the lake shore, and then through the lakes to the proximity of the iron-working region of Pennsylvania, was virtually impossible. Most of the early companies planned to smelt the ore near the mines with charcoal, and to ship pig-iron to the Pittsburgh district. The first shipments from Marquette was pig-iron, but costs of production were high and it could not compete with iron smelted in Pennsylvania.

The chief obstacle to navigation on the Great Lakes, apart from ice in winter, was the rapids on the St. Mary's river at Sault Ste Marie. In 1855 a canal was opened, circumventing the rapids, and the flow of iron-ore from Marquette to the ports on the lower Great Lakes began almost at once. Nevertheless, the idea that the shores of Lake Superior should be made to support a charcoal-iron industry was remarkably lasting. Its decline in Pennsylvania was to be attributed more to the exhaustion of the local ores than to the weaknesses in the process itself, and there were plenty of iron-masters willing and able to build charcoal-furnaces in Upper Michigan, where both ore and timber were to be had in abundance. Indeed, as late as 1903 a new charcoal-furnace was built at Marquette, and the last in this region was not blown out until 1933.

In the meanwhile, other ore ranges were discovered. The Menominee Range was opened up in the 1860's; the Vermilion in the 1870's, and the Gogebic in the 1880's. Although these ranges were at first worked from open pits, it soon became necessary to sink shafts to reach the ore-bodies. Thus mining costs gradually rose, but in 1889 a rich outcrop was found on the Mesabi Range. It was softer and more easily worked than any of the others, and the great areal extent and shallow depth of the deposit allowed it to be dug from vast open pits. In 1892, the first consignment of Mesabi ore was shipped from Duluth to the lower lakes. By 1943 it was producing 63 per cent of the total United States output of ore, and 78 per cent of the Lake Superior production.

At first the ore was shipped down through the lakes by general freighters, whose masters objected to handling the dirty cargo and took it only for lack of a better. In 1867 the Cleveland Iron Mining Co.[12] began to build its own ore boats. As the many small iron-mining concerns gradually merged to make large, stable and well-capitalized concerns, these also began to build their own ore fleets. Then industrialists, like Carnegie and Rockefeller, built their own ore boats or got possession of ore-mining companies that had their own. Today all the ore-carrying ships are owned by the mining companies, and almost all the mining companies are subsidiaries of the steel corporations.

The ships have grown from small, wooden, sail-driven craft, through iron-built ships and whale-backs, to the present-day ore-carriers which can handle up to 15,000 tons of ore. The expansion of loading and unloading equipment has kept pace with the growth of the ships. The hatches of the boats are tailored to fit the ore shutes at the loading docks, and the electrically driven Hulett unloading machines are adjusted to the size of the hatches. A Great Lakes steamer, carrying 15,000 tons of ore, can be loaded at Duluth in under three hours and unloaded at Erie or Conneaut in four hours.

The open season on the Great Lakes lasts approximately from April to December, and during this period a boat may be expected to make 30-40 round-trips. During the months when ice precludes navigation, the boats remain in dock, usually at their southern terminus, and are overhauled in preparation for their eight-month period of intensive service.

Lake Superior lies twenty-one feet above the level of the lower lakes, and the connecting St. Mary's river drops by means of rapids which have never been navigable. The first canal was built in 1855. As the size of the lake steamers increased, the original lock was enlarged and fresh locks were added. Today there are four locks on the United States side of the river and a single small lock on the Canadian. All are used intensively. Throughout the open season there is a procession of boats, almost all of them ore-carriers, through the sixty-three-mile-long St. Mary's river and the 'Soo' locks.

The lesser iron-ore ranges of Lake Superior are nearing exhaustion, though production from an increasing depth can be expected to continue for many years. Though output from Mesabi continues high the end of the high-grade ores is in sight. The future seemed bleak indeed for the smelting industry that had been geared, both by its location and its technology, to their use. It is being rescued by the development of techniques which permit the use of the more abundant low-grade ore, or taconite, in the Mesabi and other ranges, and, secondly, by the opening up of other ore-bodies accessible from the Great Lakes system (*see* below, page 137). The nearest of these is the Steep Rock ore-body which lies in the Canadian Shield north-west of Lake Superior. More extensive appear to be the Labrador deposits, which can now be used to supply the smelting industry of Pennsylvania, Ohio and the Great Lakes region without navigational problems since the opening of the St. Lawrence Seaway in 1959.   ·

The use of Lake Superior ore has had a profound influence on the location of iron and steel works. A few works had been located on the shores of the Great Lakes in order to use the small local deposits of bog-ore, but these had no lasting importance. It was the availability of Lake Superior ore, coupled with the introduction of the converter and open-hearth and the growing demand of the western markets, that brought about the industrial concentrations at Buffalo, Cleveland and, above all, Chicago. The continued prosperity and growth of these industrial regions is dependent upon finding a satisfactory substitute for the waning resources of Mesabi ore.

The Regional Distribution of the United States Iron and Steel Industry: Fig. 7 shows the distribution of iron- and steel-working in the United States in 1956. Steel-making is somewhat more widespread than iron-smelting. All the large iron-smelting centres and most of the small also have steel-furnaces, but in addition there are many small steel works not integrated with iron-working. The output of pig-iron and ingot steel by states was, in 1966:

	Pig-iron	*Steel*
New York	6,488	7,725
Pennsylvania	21,675	32,122
Ohio	16,406	22,984
Indiana	11,955	18,044
Illinois	6,639	10,960
Others	28,337	42,266
	91,500	134,101

(*Thousands of net tons*)

(Source: American Iron and Steel Institute)

Eastern District: This, the earliest home of the modern iron and steel industry in the United States, declined in relative, if not also in absolute, importance, after anthracite had ceased to be the most important metallurgical fuel. Local ore deposits, never large, were nearing exhaustion, and the most advantageous position from the point of view of the market was ceasing to be the eastern seaboard. Yet it has been the ore question that has contributed most to the present revival in the importance of this region.

The Bethlehem Steel Corporation, established in 1904 around the nucleus of an old iron and steel company at Bethlehem, Pennsylvania, came too late into the field to share in the great partition of the ore-fields of Minnesota.[13] When it also began to plan the construction of a plant on the largest and most modern scale, it had to turn to alternative sources of supply. The Corporation got possession

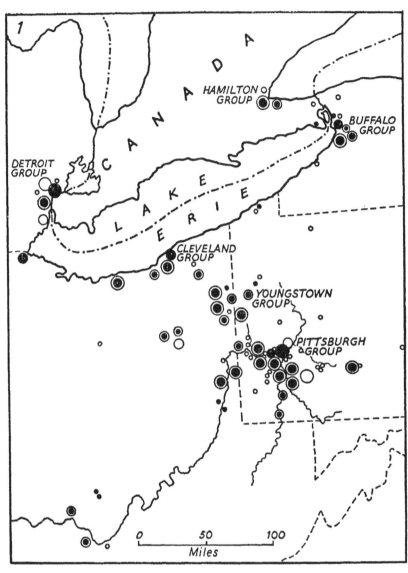

FIG. 7. Iron and steel works in north-eastern

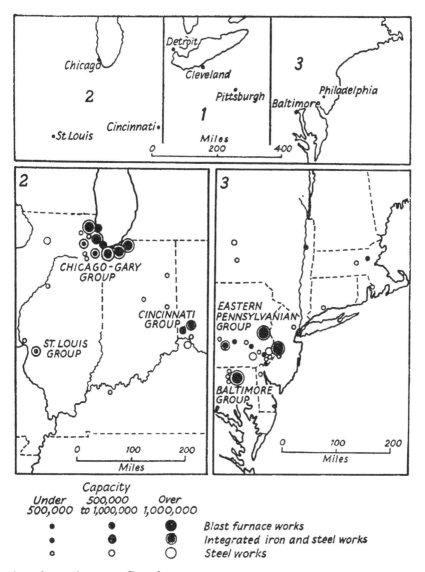

Detroit

Chicago

Cleveland

2

Pittsburgh

Cincinnati

St Louis

3

Baltimore

Philadelphia

1

Miles
200

0

400

2

CHICAGO-GARY
GROUP

CINCINNATI
GROUP

ST. LOUIS
GROUP

0   100   200
Miles

3

EASTERN
PENNSYLVANIAN
GROUP

BALTIMORE
GROUP

0   100   200
Miles

Capacity
Under   500,000   Over
500,000   to 1,000,000   1,000,000

Blast furnace works
Integrated iron and steel works
Steel works

America and eastern Canada

of an old works at Sparrow's Point, at the head of the Chesapeake Bay, in Maryland, and located its new works on this site. In 1912 it leased the El Tofo iron mines in Chile, and has since supplied its new works in the main with ore imported from this source. In recent years the Fairless iron and steel works were built on the Delaware river, near Philadelphia, by the United States Steel Corporation.

The Eastern iron- and steel-producing region includes also a group of works near Buffalo, of which Lackawanna is by far the largest. These works have the advantage of a lakeside position. Fuel is more expensive than in Pittsburgh, but the cost of ore, especially after the completion of the St. Lawrence Seaway, will be lower.

There are also a number of works in this area which derive from the earlier, anthracite-smelting, period. Foremost among these is at Bethlehem itself, where the works of the Bethlehem Steel Corporation have now a continuous history of over a century. The Johnstown works, situated in a deep valley of the Allegheny Plateau, to the east of Pittsburgh, have also grown from a small nineteenth-century works which grew up using local materials. Other smelting plants in this region are small, generally with only one or two furnaces, and, in some instances, no attached steel works.

Pittsburgh-Youngstown: This district retains the ascendancy which it gained in the second half of the nineteenth century. The impetus of its early growth, the immense capital assets which it possesses, its proximity to the market of north-eastern United States and its resources in coking coal are likely to perpetuate its dominant position. Only the problem of iron-ore supply constitutes a threat to it. For over a century Lake Superior ore has been carried by rail from the Lake Erie ports to the industrial area, and coke and, later, coking coal have been distributed from the Connellsville area. As long as a high-grade ore could be brought from the upper Great Lakes to the Lake Erie ports and overland by rail, Pittsburgh and its surrounding valleys had a price advantage over other regions. The Youngstown area, not accessible by barge from Connellsville, is at a certain disadvantage, and this has been reflected in the need to concentrate there on the production of more specialized and more highly priced steel goods that are in less direct competition with Pittsburgh itself.[14] Even Steubenville and other centres farther down the Ohio are at some disadvantage as a result of the longer haul of materials. But any increase in the cost of ore, unloaded at the lake-side docks, would be a very serious matter, and it has been fear of this that has led in recent years to the expansion of works on the Atlantic seaboard. This danger appears, however, to have been averted, and for the foreseeable future the Great Lakes and St. Law-rence freighters will be able to bring their cargoes of ore to Erie, Conneaut and Ashtabula at a cost no higher relatively than at present. The expansion of markets in the West may attract more of the industry in that direction, and strategic necessities may lead to a greater degree of dispersion of the industry than exists at present, but a failure of the supply of materials is not likely in itself to bring about a decline of iron and steel production in western Pennsyl-vania, West Virginia and neighbouring parts of Ohio, though mar-keting costs may do so.

Cleveland-Detroit: Smelting works were established on the shores of Lake Erie quite early in the nineteenth century in order to smelt the bog-ores that were found locally. But these were soon exhausted, and the modern smelting and steel-making industry came only after the Great Lakes traffic in iron-ore had been established. The earliest works were in Cleveland, first centre of the ore trade, and Cleveland remains the most important of the lakeshore sites. Lorain and Toledo still have smelting and steel works, but others established in this region have been closed. Detroit has become an important centre of smelting and steel-making, following the rise of the auto-mobile industry. There are two large blast-furnace works in the greater Detroit area, one of which is owned by the Ford Motor Company.

Chicago District: Much of the United States Industry is oriented towards a western market. The earliest iron works in Wheeling, Pittsburgh and Detroit had as their main purpose the equipping of the pioneer for his long journey to the West. Each place was the starting point of one of the routes across the Midwest. Chicago was no different. It was the starting point of the western railroads, and amongst its earliest products were iron rails. The earliest works were along the banks of the Chicago river, but the introduction of modern steel-making methods necessitated more space than was available in the heart of a rapidly growing city, and in 1880 the first works were established in the present Calumet district, sixteen miles south of the 'Loop'.[15] The works were planned to use ore from

Lake Superior, and fuel from the coal-fields of Indiana and Illinois. The latter has not proved to be wholly satisfactory, and is now replaced by coking coal from West Virginia.

A waterfront location was chosen for all works. The earliest was the Illinois Steel Co., now a part of the United Steel Corporation, with its works on the lake front, at the mouth of the Calumet river Within a few years the Iroquois Iron Co., now a part of the Youngstown Sheet and Tube Co., was established on the other side of the Calumet river. Three works were founded a short distance up the Calumet river on sites that were a great deal less convenient, with smaller scope for expansion. But the biggest development took place in northern Indiana. Here the United States Steel Corporation in 1906 bought 6,000 acres of land lying at the head of Lake Michigan to develop a large, integrated iron and steel works. At about the same time the Inland Steel Corporation bought a tract of land at Indiana Harbor, described at this time as the nearest stretch of lake frontage to Chicago that had still not been built up.

The land around the head of Lake Michigan was of no agricultural value. Low-lying and swampy, it could be used for housing only with difficulty. But all the main railroads from Chicago to the East ran past the area—and in some instances through it. Dock basins could be excavated in the soft glacial deposits, and land recovered from the lake floor by emptying there the unwanted slag from the blast-furnaces and steel works. It was here that the United States Steel Corporation built the city of Gary[16] and the works of its subsidiary, the Indiana Steel Company. The city was well planned and is one of the less unattractive of the country's steel cities. The works, expanding over land built up from the lake floor, have long been regarded as the largest in the world, with twelve blast-furnaces, a pig-iron capacity of about 5 million tons a year and a steel capacity of nearly 8 million tons. Only very recently has the capacity of the Sparrow's Point plant exceeded that of Gary. To the west, at Indiana Harbor, are the works of the Inland Steel Corporation and the steel works of the Youngstown Sheet and Tube Co., and to the east of Gary other steel works are now being built.

The Indiana-Illinois region has every advantage except that of fuel supply. In particular, it has what the older centres of Pennsylvania and Ohio lack, an abundance of level land over which the plant can spread in whatever manner is best suited to its own internal organization. It is a comparatively low-cost producer, and the abolition of the basing-point system of assessing prices has increased its importance in the Western market.

The Chicago district also includes works at St. Louis, Kansas City and Duluth, Minnesota. The last is small. Its advantage lies in proximity to the Mesabi ore deposits, but fuel is expensive, markets distant, and conditions of life almost as bleak as at the ore-mines themselves.[17]

Southern District: In the middle years of the nineteenth century the iron industry was spread thinly through the American South.[18] Iron-ores were abundant but of low grade; charcoal was plentiful, and the scattered furnaces met the needs of small and often isolated communities. It was the most primitive branch of the American industry, and was protected from the competition of the North only by its distance. It was, nevertheless, in steady decline after the Civil War, when the developing railroad system made possible the distribution of the cheaper products of Pittsburgh and the Lehigh Valley.

But there were exceptions to this picture of a general decline. Coking coal was found in Tennessee and was used to smelt the red iron-ores of the southern Appalachian mountains. Chattanooga became a centre of the smelting, refining and rolling industries. But its prosperity was short-lived, and it succumbed to the competition of the better-placed industry of the Birmingham district of Alabama.

Birmingham is often quoted as the ideal site for an iron industry, with all the raw materials to be found within a mile or two of one another. 'The Birmingham companies are perhaps the only ones in the United States where the general management can reach all important operating mines and plants by local telephone or can at any time visit personally any unit in a matter of minutes.'[19] This undoubtedly oversimplifies the situation at Birmingham. The red ore bed of Birmingham is part of the folded Appalachian mountains, and stretches as a long narrow belt (Fig. 8) from north-east to south-west through the northern half of the State. The ore outcrops, and could, at least in the early years of its exploitation, be mined comparatively easily. Mining costs have now gone up sharply; the ores are low in iron content, and have a high percentage of silica and phosphorus. They are amongst the more expensive ores to mine. The coal of the Warrior and Coosa basins lies at a higher horizon than the ores and in part overlies them, and much of the Warrior basin coal is of good coking quality. Between the outcrop of the red ore and the Warrior basin are extensive deposits of good, fluxing limestone.

It was not until about 1880 that this close juxtaposition of the raw materials was used to establish an iron-smelting industry. Though iron could be produced here cheaply, the South at this date was unindustrialized, had only a low purchasing power and could not provide a market for large quantities of varied iron and steel goods. Indeed, cast-iron pipes for sewers and water-supply seemed to be in greatest demand. In 1883 the manufacture of pipes was established in Birmingham, and within a few years the city was producing a major part of the country's output of iron pipes.

Pig-iron smelted from the red ore was highly suited to casting, but had too much silica for the basic open-hearth process and not enough phosphorus for the Thomas converter. It was only after many years of experiment that a low-silica pig, suitable for the open-hearth, was made at Birmingham. In the North the open-hearth process was operated with a large (45–55 per cent) proportion of scrap. But the South was not in a position to provide the scrap, without which the open-hearth process is relatively expensive. The solution here was the adoption of the Duplex method, by which the converter is used to burn off the carbon and silica, while the metal is finished off, with the elimination of the phosphorus, in the open-hearth. This is now the standard practice in at least one of the works of this area.

The smelting industry disappeared from most of the South, but has survived also in Anniston and Gadsden, small cities within sixty miles of Birmingham. Control of the industry and ownership of its local resources has now come to be concentrated in only four concerns, the largest of which is a subsidiary of the United States Steel Corporation, and the second largest, a division of the Republic Steel Corporation. The operating companies own, between them, all the available local resources for the industry. In a nation which sets a high value on private enterprise and individual initiative, this monopolistic position is not viewed with complete equanimity. One of the four companies is known to be in a very weak position as regards its ore supply, and has no opportunity to extend its holding

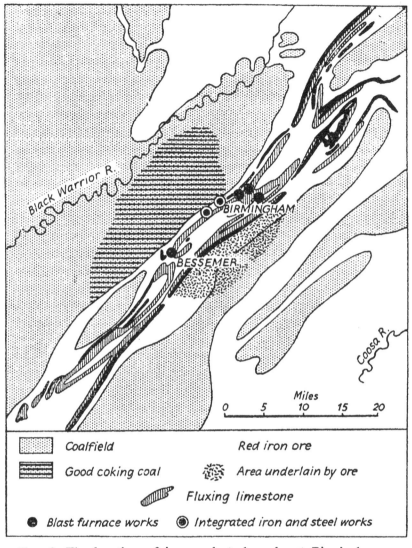

FIG. 8. The location of iron and steel works at Birmingham, Alabama

of ore deposits. Furthermore, 'The complete ownership by the operating companies of the known reserves of commercial red ore in the district tends to preclude an expansion of operations through the establishment of new companies.'[20]

Western District: In recent years a smelting and steel-making industry has grown up in the West. The oldest centres are at Pueblo, Colorado, and Ironton, Utah. At Geneva, Utah, is a newly built integrated works. But the greatly increased demand for steel goods that arose just before and during the Second World War led to the establishment of iron- and steel-furnaces at Fontana, California, and at Houston and other points in Texas. The West constitutes a large market, but it is one that is thinly spread over a very large area. It is doubtful whether any site within the West has any marked advantage in distribution costs over the Chicago area, though one hears of plans to build a plant west of the Mississippi. On the other hand, the West has sizeable deposits of ore, and its supplies of coking coal are at present adequate. In general, the higher production costs are not offset by lower distribution charges, and one should probably not look for any conspicuous change in the present geographical pattern of iron and steel production in the United States.

Canada: Canada, like the United States, had a rudimentary iron industry soon after its first settlement by European peoples, but, unlike the American industry, it failed to develop beyond the charcoal stage.[21] The last of the primitive furnaces was extinguished in 1883. In the meanwhile, however, a puddling- and later a steel-making- and rolling-mill was established in Nova Scotia for the purpose of making rails. It was not, however, until about 1900 that a modern blast-furnace works and integrated steel works was set up.

The slowness of industrial development in Canada, at least in comparison with that of her southern neighbour, is to be attributed partly to lack of fuel resources, but mainly to the smallness of the local market. The industry grew exceptionally slowly, and before the outbreak of the Second World War production of pig-iron reached a million tons only in a few boom years. Steel production was generally somewhat larger; it began regularly to exceed a million tons only in 1936. During and since the war, however, economic growth in Canada has achieved more significant proportions, and with it the iron and steel industry has grown more rapidly.

Production of pig-iron rose from 1·3 million tons in 1940 to 5·49 in 1963, while the growth of steel output rose from 2·07 million tons to 7·43 in the same period. A greatly increased demand for steel goods is anticipated, and existing works are being expanded.

Canada is undoubtedly rich in reserves of high-grade iron-ore, but these are in general difficult of access and their exploitation is only now beginning. Foremost among these are the deposits which lie on the borders of Labrador and Quebec. Their discovery is comparatively recent, and the Iron Ore Company of Canada was not formed to exploit them until 1949. Ore-bodies found thus far form a belt, some ninety miles from north-east to south-west and perhaps not more than a tenth of this in width. It is, however, far too early to make any pronouncement regarding the extent of the deposits, and they may well prove to be far more extensive than those of the Lake Superior region. A railway has been built from Seven Islands, on the shore of the Gulf of St. Lawrence, northwards into the Shield. The ore is being worked in large open pits and is already being shipped from Seven Islands in the open season.[22] Conditions of ore-mining and transport here are fair more severe than in Minnesota. The winters of exceptional severty, the heavy snowfall and the difficulty of keeping the railway open will all add to the cost of the ore.

The exploitation of the second major source of ore in Canada is also beset by technical difficulties. The Steep Rock deposits constitute an ore-body similar to the Mesabi, but it lies to the north-east of the latter and inland from Port Arthur and Fort William. Furthermore, it underlay the Steep Rock lake, which has had to be drained and rivers diverted. The mines came into production in 1953, and the deposit is well on the way towards replacing the Mesabi as a major source of ore for Canada.[23]

The third important source of ore in Canada is Bell Island, in Conception Bay in south-eastern Newfoundland. This, the most accessible of the larger deposits, has been used for many years to supply the furnaces at Sydney, Nova Scotia. Some small deposits of ore in Quebec and Ontario have been exhausted; others on the Shield and in the Rocky Mountains are barely known, least of all opened up.

In contrast with its abundance of ore, Canada is sadly deficient in coking coal, and the only significant deposit is that at Sydney on Cape Breton Island. This coal is adequate in quality, but is becoming increasingly difficult and expensive to mine.[24] The coal of Cape Breton Island is taken by boat up the St. Lawrence river to the iron-smelting centres in Ontario, but it is a very expensive fuel when it reaches its destination, and fuel from the Appalachian coal-field of the United States is generally preferred.

The iron and steel industry of Canada is principally divided between three centres: Sydney (Nova Scotia), Hamilton and Welland, and Sault Ste. Marie (all in Ontario). Sydney is the oldest centre and is the only one that is well located in relation to all its raw materials. It uses coking coal from the Cape Breton coal-field and iron-ore from Bell Isle. It is, on the other hand, relatively remote from its markets, and it is the least important of the centres of production, and the one which is expanding the least rapidly.[25]

Over 50 per cent of Canada's iron- and steel-making capacity is at Hamilton, Ontario. The situation at the head of Lake Ontario is well suited to the transport of ore by water either from the St. Lawrence or from the Upper Lakes, and fuel is obtained mainly from the Appalachian coal-field, which is no more than 250 miles distant. There are two integrated iron and steel plants in Hamilton and another at Port Colborne, on the Welland Canal, only forty miles away. In addition there are two steel-making works in the same area. The total capacity of the Hamilton-Welland region is 1,772,800 tons of pig-iron and 2,968,800 tons of steel. It lies in the heart of the manufacturing belt of Canada, close to the United States' steel-making centres of Buffalo and Cleveland, and is as well situated as any place in Canada in relation to the sources of materials that are at present known.[26]

The third important centre of the iron and steel industry is at Sault Ste. Marie, on the northern shore of the St. Mary's river which joins Lake Superior to Lake Huron. It was originally established to smelt the ore from the Helen mine, a short distance away to the north-west. Limestone was obtainable locally, but fuel had to be brought in from Pennsylvania and Ohio. The local ore-body has been exhausted, and Mesabi and Steep Rock ore supply the Algoma works. Sault Ste. Marie is within the boundary of the Canadian Shield and marginal to the industrial markets of Ontario and Quebec. Its relative distance from sources of coking coal and metal scrap may not be wholly compensated by its closer proximity to the expanding markets of the West.

These three centres account for all of the Canadian pig-iron production and 95 per cent of the steel. The remaining steel producers are made up of four works which produce mainly electric steel. Their situation has been determined very largely by the availability and price of the electric power which this process requires in such large quantities.

Present plans are for a considerable increase in the steel-making capacity of Canada, including an increase of the manufacture of tubes, so much needed in the growing petroleum industry. It is significant that all the published plans call for increases to the plant in the Hamilton district. The Canadian output of iron-ore has increased very rapidly in recent years, from 1,456,000 tons of metal content in 1948 to 21,658,000 in 1965. This total may be expected to grow as the Steep Rock and Labrador ore-fields are opened up.

[1] *Resources for Freedom* (Report of the President's Materials Policy Commission), Government Printing Office, 1952, II, 11.
[2] J. M. Swank, *History of the Manufacture of Iron in All Ages*, Philadelphia, 1892.
[3] N. Hartley, *Ironworks on the Saugus*, University of Oklahoma Press, 1957.
[4] J. M. Swank, *op. cit.*, 120–35.
[5] Charles S. Boyer, *Early Forges and Furnaces in New Jersey*, Philadelphia, 1931.
[6] J. P. Lesley, *A Collection of Occasional Surveys of Iron, Coal and Oil Districts in the United States*, Philadelphia, 1874.
[7] James H. Bridge, *The Inside History of the Carnegie Steel Company*, New York, 1903. *See also* J. P. Lesley, *The Iron Manufacturer's Guide to the Furnaces, Forges and Rolling Mills of the United States*. New York, 1859.
[8] H. B. Vanderblue and W. L. Crum, *The Iron Industry in Prosperity and Depression*, Chicago, 1927, 62ff.
[9] L. White, 'The Iron and Steel Industry of the Pittsburgh District', *Economic Geography*, IV, 1928, 115–39.
[10] Allan Rodgers, 'Industrial Inertia—a major factor in the Location of the Steel Industry in the United States', *Geographical Review*, XLII, 1952, 56–66. *See also* the successive issues of the *Directory to the Iron and Steel Works o the United States*, Philadelphia, from 1873.

[11] Harlan Hatcher, *A Century of Iron and Men*, Indianapolis, 1950. For the interesting legal aspects of this development, *see* Fremont P. Wirth, *The Discovery and Exploitation of the Minnesota Iron Lands*, Cedar Rapids, Iowa, 1937.

[12] Later absorbed into the Cleveland Cliffs Corp.

[13] Arundel Cotter, *The Story of Bethlehem Steel*, New York, 1916.

[14] Allan Rodgers, 'The Iron and Steel Industry of the Mahoning and Shanango Valleys', *Economic Geography* XXVIII, 1952, 331–42.

[15] John B. Appleton, 'The Iron and Steel Industry of the Calumet District', *University of Illinois Studies in the Social Sciences*, XIII, No. 2, 1925.

[16] Judge Elbert H. Gary was at the time President of United States Steel Corporation.

[17] Langdon White and George Primmer, 'The Iron and Steel Industry of Duluth: a Study in Locational Maladjustment', *Geographical Review*, XXVII, 1937, 82–91.

[18] H. H. Chapman, W. M. Adamson, H. D. Bonham, H. D. Pallister and E. C. Wright, *The Iron and Steel Industries of the South*, Bureau of Business Research, No. 17, University of Alabama, 1953.

[19] *Ibid.*, 155.

[20] *Ibid.*, 155.

[21] J. W. Swank, *op. cit.*, 348–51.

[22] *The Canadian Mining and Metallurgical Bulletin*, LI, 1958, 505–11.

[23] *New York Times*, 26 September, 1957.

[24] A. W. Currie, *Economic Geography of Canada*, 71–77.

[25] A. W. Currie, *op. cit.*, 75–76.

[26] 'Steel in Canada', *International Iron and Steel*, December 1956, United States Department of Commerce; Donald Kerr, 'The Geography of the Canadian Iron and Steel Industry', *Economic Geography*, XXXV, 1959, 151–63.

7

THE IRON AND STEEL INDUSTRY
OF THE SOVIET SPHERE

THE heavy industries play a big role in Communist ideology, in part because their development is of great military, economic and social importance in these predominantly agricultural countries; in part because it was in the factory workers that Communism has always found its most disciplined supporters. Within the Soviet Sphere, the iron and steel industries have grown in recent years at an accelerating pace, and the plans that have been devised for them anticipate a yet more rapid rate of growth. But in their earlier history, and in their current technology, these industries belong to the West. The knowledge of technical processes has spread from Western Europe to Eastern and then to Russia. Technicians— British, French and German for the greater part—were at intervals, from the late eighteenth century onwards, brought in to build up works and to teach their skills to the local peoples.

Eastern Europe and Russia have not merely borrowed the skills evolved in the West. They used them long before the Communist Revolution to build up an iron industry on a massive scale. In the late eighteenth century, Tsarist Russia became one of the greater producers of pig-iron and by far the largest exporter. There was a time when much of Sheffield's steel was refined from pig-iron smelted in the Ural mountains. It was within the borders of present-day Poland that coke was for the first time in continental Europe used successfully in the blast-furnace. The former Austrian lands in Eastern Europe were quick to adopt the technological advances made in the nineteenth century, and in not a few respects the processes were improved upon in these countries.

There is, then, nothing particularly new in the metallurgical developments of the countries of the Soviet Sphere, with the exception of China. The foundations of the modern industry were well and truly laid a century ago. On these foundations the Communist planners have built at a fast—at times reckless—speed. In this chapter, the iron and steel industries of the Soviet Sphere are divided, for reasons that will become apparent, into three sections: the industries of Communist East-Central Europe; of the Soviet Union itself, and of Communist China.

East-Central Europe: The six countries[1] which make up East-Central Europe, together with the Eastern Zone of Germany, have each an iron and steel industry that was developed before the Second World War, and in most cases before the First. The region as a whole is poorly endowed with resources. Iron-ore deposits are numerous but small, too small in most instances for economic exploitation. Although East-Central Europe contains, in the Silesian-Moravian coal-field, one of the largest coal basins in the world, the amount of coking coal is severely limited. The other coal basins of the area are small in total resources and in no instance do they produce a satisfactory metallurgical fuel.

In spite of these considerable handicaps, the metallurgical industry was developed widely in the East European countries in the nineteenth century. The scale of the units established was generally small, so that the slight reserves of materials could be exploited effectively. The poorly developed means of transport prevented really effective competition from the larger and more economic units of production in Western Europe, and the local market among the peasant peoples of Eastern Europe was itself growing but slowly.

As a general rule, the plant established at this time in Eastern Europe was equipped and staffed with Western capital and Western technicians. In the territories of present-day Poland and Czechoslovakia this capital was mainly German, though in those parts which belonged to the Russian Empire French capital became preponderant at the end of the nineteenth century. In the Danubian lands Austrian capital was more conspicuous. By and large, these foreign investments in the heavy industry of Eastern Europe survived the boundary changes and the creation of new political units that followed the First World War. The French share in the

FIG. 9. Iron and steel works in Communist Eastern Europe
The Košice works are projected, but not built at the time of writing

investment was intensified, especially in Czechoslovakia, where the French firm of Schneider Frères formed a close association with Škoda, and in Poland. German holdings in Poland were retained, and elsewhere in Eastern Europe German participation in industry was revived.

During the years of German occupation of this area in the Second World War, all plant and equipment that could be of service to Germany were taken over and, in most cases, entrusted to one of the big German steel concerns. In a few instances the plant was extended, but in general it was ruthlessly exploited, and left in 1945 in a very poor condition. The rehabilitation of these countries after the German defeat and withdrawal was controlled by a series of national plans. After their completion a further series of plans aimed at increasing the capital investment and the productivity of each of the countries of Eastern Europe.[2]

In the seven countries of Eastern Europe production of both iron and steel was in 1957 about three times the pre-war level. Only Poland increased her capacity as a result of territorial changes. In all countries existing capacity has been more fully used, and, as will be shown, new capacity was built on a lavish scale.

	Pig-iron	Production	Steel	
	(thousands of metric tons)			
	1938	1965	1938	1965
Poland	879	5,760	1,441	9,088
Czechoslovakia	1,323	5,927	1,873	8,598
Hungary	335	1,588	648	2,520
Yugoslavia	75	1,176	227	1,769
Romania	123·7	2,019	276·5	3,426
Bulgaria	—	695	—	588
East Germany	74	2,338	1,695	3,890
TOTAL	2,809·7	19,503	6,160·5	29,879

Poland: The territorial changes which followed the defeat of Germany gave Poland control over about three-quarters of the Upper Silesian-Moravian coal basin as well as of the small Walbrzych field in Lower Silesia, valuable for its coking coal.[3] To her own industrial plant in Upper Silesia she added the smaller capacity of West or German Upper Silesia. The result was the formation of an industrial complex, somewhat resembling the Ruhr, though appreciably smaller in scale.[4] In the triangle formed by the cities of Gliwice, Bytom and Dąbrowa Górnicza there were six integrated iron and steel works, a blast-furnace works and six steel-making- and rolling-mills. Their total capacity was, however, no no more than about 1·7 million tons of pig-iron and 2·0 million tons of steel. The works were small, and some were old and poorly maintained. Reasons for the small scale can be found both in the political uncertainty which has long clouded the future of this area and also in the technological conditions imposed by the local fuel supply. Upper Silesian coal yields a friable coke which could be used only in small stacks.

In recent years plant has been modernized and extended, notably at the Kościuszko and Pokój works, but the iron and steel industry labours under a grave handicap. Apart from the shortcomings of the coal supply, the local reserves of iron-ore are for practical purposes exhausted, and there is little chance of replacing them from other parts of Poland. At the present time the ore-mines of Krivoi Rog in Ukraine S.S.R. are the main source of supply to Polish furnaces, supplemented by the domestic *minette*-type ores and by imports from overseas. Furthermore, the situation of the industrial area on the divide between the basins of the Odra and Vistula makes water transport impossible within and close to the industrial area. It is probably in this respect that Upper Silesia suffers most in comparison with the Ruhr.

In spite of the poverty of natural resources, Poland has extended her iron and steel industry since 1950 by the erection of three new works. Foremost among these is the Lenin works at Nowa Huta, built, with its accompanying city for almost 100,000 persons, near the Vistula to the east of Kraków. It is a fully integrated works, and when completed will have a steel capacity of about 1·5 million tons. Its continuous plate-mill is said to be the first ever to be built by the Russians; it is not easy to understand why they should have allowed it to be established in Poland.[5]

The second works is the Bierut plant, another fully integrated undertaking, though smaller than the Lenin works. It lies at Częstochowa, near the foot of the limestone, ore-bearing scarp of the

Krakowska Jura. The location is very similar to that of the Lorraine works; unfortunately, the ore reserves of the Jura are in no way commensurate with those of *minette*. The third plant is a steel and rolling-works in Warsaw. Poland has also some small works in the south-east of the country.

It seems likely that the fuel difficulty may largely be overcome as methods are devised of producing metallurgical coke from the Upper Silesian coal. But the shortage of ore can only become more acute as the industry expands. It is reported that deliveries of Ukraine ore have been adequate, and also that output at Krivoi Rog continues to increase, but this situation clearly depends on Russian goodwill. Poland is known to be interested in the ore-bodies at present being opened up in the New World.

Czechoslovakia: In the past Czechoslovakia has been somewhat more developed industrially than Poland, and its steel industry, in consequence, larger and more varied. But her resources are no greater. Her share in the Upper Silesian-Moravian coal-field is small, though the proportion of coal that is of coking quality is much higher than in Poland. The small coal basins of Bohemia do not yield coking coal, and the iron-ore of Bohemia and eastern Slovakia is of good quality but small in amount.[8]

The centre of the Czechoslovak iron and steel industry lies in northern Moravia, where the important works of Vítkovice were established early in the nineteenth century. The second large works is at Třiniec in the nearby territory of Těšín (Teschen). To these two has been added a new works at Kuncice, near Moravska Ostrava, known as the Gotwald Huta. All three are integrated works, using local coal and ore brought in by rail from Slovakia, Bohemia and also from the Soviet Union.

There are a number of older and smaller works in Bohemia. The only integrated works is at Kladno, but there are blast-furnaces at Beroun, and steel-making and rolling works at Plzeň (the Škoda concern), at Chomutov, where Mannesmann established a tube-mill in the nineteenth century, at Most and at Prague. An integrated works has now been built near Košice, in the far east of the country, in a formerly backward area which it is sought to industrialize.

Like Poland, Czechoslovakia plans to extend her iron and steel industries, but, also like Poland, she faces a serious shortage of ore

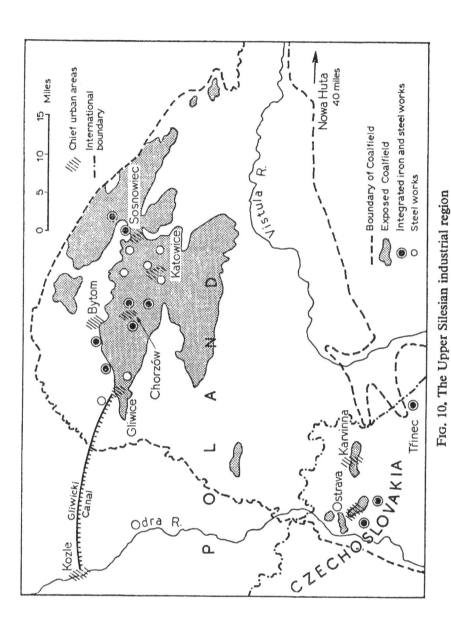

FIG. 10. The Upper Silesian industrial region

The Gliwicki Canal, from Gliwice to the Odra river, replaced by the old Kłodnicki Canal

and her problems of transporting raw materials are no easier than those facing Poland.

Hungary: There has long been a smelting industry, based upon the ore and charcoal of the hills along the Slovakian border. In recent years these works, chiefly in and near Miskolc, have been modernized and extended. Modern Hungary, however, has little iron-ore, though the small Pécs coal-field yields coking coal in quantities which are said to be adequate for the Hungarian industry. A new, integrated works has recently been built at Dunapentele, on the Danube some thirty-five miles below Budapest. These works began production in 1954, though construction has not yet been completed. Hungary appears to rely in part on Russian ore, brought up the Danube from the Black Sea.

Yugoslavia: Iron-ore resources are amongst the largest in Eastern Europe, but the coal is poor in quality and inadequate in quantity. A small iron and steel industry was developed, largely by Austrian initiative, in the nineteenth century, but fuel was obtained mainly from the West. No doubt it was planned that the newly developed Yugoslav industry would derive its metallurgical fuel from Czechoslovakia. The breach between Tito and the *Cominform* has jeopardized this supply, and in recent years Yugoslavia has been importing coking coal from the Free World. Some success, however, is claimed for a process that has been developed to use the low-quality domestic coal.[7]

Since the war some of the older and smaller centres of production have been closed, and the greater part of the Yugoslav production is now from three small, integrated works at Jesenice in Carinthia, at Zenice-Vareš in Bosnia and at Sisak, near Zagreb. This last is a new plant, and other works are projected, or being built, at Sarajevo, Skoplje and Titograd. The works, both projected and already existing, are close to ore deposits but, as already noted, their fuel supply is precarious.

Romania: The iron and steel industry of Romania is younger and less developed than that of the countries of Eastern Europe discussed thus far.[8] Reserves of both ore and of coking coal appear to be too small and, by themselves, would not justify the erection of a large plant. The Romanians are at present building a large, integrated plant at Galați, near the head of the Danube delta. The chief centres of production are the old works at Resita and Hunedoara, where some modernization of very old equipment has taken place.

Bulgaria: The known reserves of coking coal and of iron-ore are very small. There had been for many years a small charcoal-iron industry. Then lignitic coke began to be used, and in recent years two integrated works, the first at Pernik, the second at Kremikovci, near Sofia, have been built, and operate in part on imported materials.

East Germany: The territory of the East German Republic was, before the war, more important for the fabrication of steel goods than for the production of iron and steel.[9] Resources in coking coal and ore are small, though probably greater than those of any of the Balkan countries. The only integrated works that the Republic inherited from Hitler's *Reich* was the old and rather small Maxhütte, at Unterwellenborn in the Thuringian mountains. There was, however, a number of steel-making and rolling works. During the past decade a smelting plant has been built at Calbe, and an integrated works on the Oder, near Frankfurt. It appears to have been planned to use Silesian coal and ore imported probably through the port of Szczecin and brought up river by barge.

Seven countries have been very briefly discussed. In two of them, Poland and Czechoslovakia, the current expansion of the iron and steel industries is justified by their long tradition of iron-working, by their possession of reserves of technical skill and of metal-using industries which provide a ready-made market. Furthermore, both have reserves of some of the needed raw materials. The expansion in East Germany is, in a sense, 'unnatural', in so far as it could not have taken place if Germany were united, and the markets of East Germany open to the products of heavy industry in West Germany. Elsewhere, the expansion of smelting and steel-making has been forced through by the Communist planners in the face of a poverty of resources and a lack of metallurgical skills. The other four countries are heavily dependent on imported raw materials; the scale of production is very small and costly processes are used in order to adapt local materials to at least some of their needs. We shall return in the next chapter to this question of the 'ideological' development of the iron and steel industries.

The ancient charcoal industry of Russia was very small in scale and was carried on mainly in the regions of Moscow and Tula. At the beginning of the eighteenth century the iron industry of the Ural mountains was developed, at the behest of Peter the Great, to satisfy Russia's military needs.[10] The Central Urals were rich in ore; forests were abundant, and though the population was scanty, it could be —and was—impressed to provide labour for charcoal-burning, mining, iron-working and transport. The earliest works in this region were State-owned, but Nikita Demidov, an iron-master from Tula, was invited by Peter to establish himself there. This he did, and founded a dynasty of iron-founders that was without equal.[11]

The early works were in the central Urals. Later in the eighteenth century industry spread into the southern Urals, where it predominates today. State-owned and private establishments existed side by side; Russian craftsmen were aided by technicians introduced from the West. The volume of production grew rapidly. It sufficed for Russian needs, and despite the long and hazardous journey by boat and pack-animal, Russian iron was marketed in the Baltic ports. It came to exceed the carefully controlled Swedish export and in the later years of the century provided the bulk of Great Britain's large import of iron.[12]

The Revolutionary and Napoleonic Wars interrupted trade just at the time when Great Britain was extending her coke-smelting industry and thus needed to import less bar-iron. In other European countries the market for Russian iron disappeared, and within the Russian Empire itself a new industry, based on mineral fuel, grew up at Tula, in Poland, and in the Ukraine in the second half of the nineteenth century. In 1880 the total pig-iron production of Russia amounted to only about 450,000 tons, of which the Urals still accounted for two-thirds. Although the iron-working industry of the Ural mountains continued to expand slowly under the Tsars, its place was taken in the next few years by that of the Ukraine.

The iron-ore reserves of Russia have already been described (pages 41–43). The most valuable of them all, those of Krivoi Rog, in the Ukraine, began to be exploited in 1880 by the Welsh *émigré* iron-worker, John Hughes. The Krivoi Rog Iron Co. was founded in 1880, and in this year the iron production for the whole of the Ukraine was only 21,000 tons. In 1885 there were still only two blast-furnaces, but thereafter the growth of the Ukraine industry was rapid. Pig-iron production reached 220,000 tons by 1890, and 1,005,000 in 1898.[13] At the end of the century the Ukraine accounted for just 50 per cent of the total Russian output, the Ural mountains for about a quarter, and the remainder came from Poland, the Moscow-Tula region, Finland and Siberia. At the time of the Bolshevik Revolution the Ukraine, which in effect meant the Donbas, produced about three-quarters of the Russian output of iron and steel.

Planned Growth in the Soviet Union: It was not until 1929 that iron production again reached the level of the pre-war years. The First Five Year Plan called for an expansion of the iron and steel industry; so did the Second Five Year Plan which launched the iron and steel plants at Stalinsk in the Kuzbas, western Siberia. Up to the outbreak of the Second World War there was a fourfold increase in the production of pig-iron and an even greater expansion in steel output. In 1940 Soviet pig-iron production reached 14,902,000 tons, and crude steel 18,300,000 tons.[14]

The German invasion of the Soviet Union greatly reduced output; plants in the Ukraine, Leningrad and Stalingrad were overrun, although much equipment was removed to prepared sites in the Urals and beyond. By 1949 the level of production had again reached that of 1940, and the last ten years have been a period of continued expansion. In 1965, 66,184,000 tons of pig-iron and 91,021,000 tons of steel were produced, and the planned objectives for 1970 call for 145 million tons of steel.

This expansion of output has been accompanied by a re-location of industrial plant. The Soviet Union is a country of vast extent, and the little-known wastes of Siberia are thought to be richly endowed with minerals. The Soviet plans aimed at dispersing industry more widely and at preventing, for social as well as military reasons, an over-great concentration in the Ukraine.

The Ukraine was well placed for the industry, with the best coking coal in the Soviet Union at the time at hand and the Krivoi Rog deposits of high-grade iron-ore at no great distance. The Ural mountains were richly endowed with ore, but the coal-fields of this region were small and the coal unsuitable for metallurgical use. It was hoped to organize an exchange of Urals iron-ore for Donbas coal, but no progress was made in this direction before the Revolution. About 1,200 miles to the east of the Ural mountains, however, lies the coal-field which has come to be known as the Kuznetsk basin or Kuzbas.[15] Its reserves are very large, a small part of them being made up of good-quality coking coal. The Karaganda coal-field in the Kazakh S.S.R. is easily worked but its coal has a high sulphur and ash content, and it too is not really suited for coking purposes. The same is true of the Kizel coal basin in the northern Urals and of the East Siberian field, near Irkutsk. Ironically, it seems that the only Soviet coal basin, apart from the Donbas and Kuznetsk, that can provide coal of unquestioned coking quality is the Pechora basin, near the shore of the Arctic Ocean. Here the severe climatic conditions are said to increase the costs of coal production by 50 per cent.[16]

From the available coal resources, the Russians chose, at the beginning of their Second Five Year Plan, to stress those of the Kuznetsk basin, and to integrate them with the ores of the southern Ural mountains. A modern, integrated iron and steel works was established at Magnitogorsk, in the Urals, and another at Stalinsk, in the Kuznetsk basin. The Urals-Kuzbas combine has about it a simplicity that makes an immediate appeal. The shuttle-service of cars carrying ore in one direction, fuel in the other across the level plains of western Siberia is a bold answer to the immutable condition set by nature. The reality, however, is somewhat different. The better Magnitogorsk ores, like those of Minnesota, are nearing exhaustion, and those that remain need beneficiation before making so long a journey. Nor does it appear that the Kuznetsk coking coal is always worth the cost of its transport all the way to the Ural mountains. The transport of materials has strained the railway facilities, and the Russians did not in their later plans extend the smelting capacity in the Kuzbas. Indeed, as Professor Gardner Clark has observed, 'ever since the famous Urals-Kuzbas combine came into full operation, the Soviets have been trying desperately to break it up again'.[17]

Actually, had the coal resources of Karaganda in Kazakhstan been better appreciated at the time, a Ural-Karaganda combine might well have provided a better economic proposition than the Ural-Kuznetsk combine, their relative nearness compensating for the lower quality of Karaganda coal. Coking coals are now railed there from Karaganda and meet an increasing share of the requirements. The recently discovered ore deposits of Kustanay in northwestern Kazakhstan will probably be used to supply the Urals industry where the Magnitogorsk deposits are running out.

The Scale of Soviet Industry: Mere size always seems to have had an appeal to the Russians. In the eighteenth and nineteenth centuries some of the largest blast-furnaces were to be found in Russia, and in the present century the Russians have, more than any other people, appreciated the advantages of the large scale on which American plants operate. There are economies and advantages in operating on a large scale, as we saw above (Chapter 3). But these advantages are limited. A small number of large units makes for inflexibility. It is easy to understand how the earlier policy of building big, specialized works placed a burden on the transport system that it really could not bear, and it is noteworthy that the more recent plans have called for plant of more moderate size.

A reason for building large plant was undoubtedly that it economized in the use of the scarcest commodity in both Tsarist and Communist Russia, technically trained personnel. Unfortunately the technical problems raised by the large scale of operations often outran the technical competence of the management. This was especially so when low-grade, siliceous ores were smelted with coke, high in sulphur and ash, in blast-furnaces of exceptional size. Furthermore, the pig-iron drawn from such furnaces called for especial care in refining and rolling.

The scale of open-hearths has also been increased so that in this respect also their size is second only to that of those in the United States. The technical problems of the iron and steel industry are those which any country would face that attempted to operate on such a scale, and the difficulties which they are known to have do not detract from the technical competence of these branches of industry. The Soviet rolling industry is appreciably less advanced. The rolling-mills were formerly small, but under the Five Year Plans they are being built larger and more specialized. Even so, more blooming-mills are needed, which can reduce the large ingots poured from large steel-furnaces to the requirements of the sheet-, strip- and rail-mills. There are probably no more than three continuous hot strip-mills, and only one of these was created by the Russians themselves.[18]

The Geographical Pattern of the Soviet Industry: One of the objectives of Soviet planning was, as we have seen, a wider dispersal of the industry. This has been achieved, though with considerable difficulty. The natural resources do not lend themselves to this kind of a dispersion. Despite the devastation of the war and the efforts of the planners, the leading Tsarist area of production, the Ukraine, produced just a half of the pig-iron in 1954 and 37 per cent of the steel.[19] Its industrial region occupies a belt of country about eighty miles across and stretching some 350 miles from Krivoi Rog eastwards to the bend of the Donets river. Across this belt, near its western end, flows the Dnieper river. To the west of this, enclosed by the big bend of the Dnieper, is the Krivoi Rog ore-body, while at Nikopol on the Dnieper are rich manganese deposits. The Donets coal basin lies in the east of this belt, within the similar bend of the Donets river.

The iron and steel plant can be grouped into: (*a*) those lying along the Dnieper river, and thus relatively close to the ore, these include the large, integrated works at Dneprodzerzhinsk, Dnepropetrovsk and Zaporozh'ye, and (*b*) the works on or close to the Donets coal-field. This is the older centre of industry, and the works are somewhat smaller. The chief integrated works are at Stalino, Kramatorsk, Konstantinovka, Makeyevka and Vorozhilovsk. The plants in both groups rely almost exclusively on the ore and coking coal from the region itself. Lastly, (*c*) the use of the low-grade ores of the Kerch peninsula in the Crimea has made it desirable to establish smelting and steel works close to the ore itself. An integrated works was built at Kerch, but it was severely damaged during the war, and it is uncertain whether it has yet been completely restored. On the shore of the Sea of Azov, opposite Kerch, are two integrated works at Zhdanov.

The Urals region is second in importance only to the Donbas.[20] The region was, of course, the scene of the eighteenth-century industrial expansion, and, though the area lost relatively during the later years of the nineteenth century, it has never ceased to be a

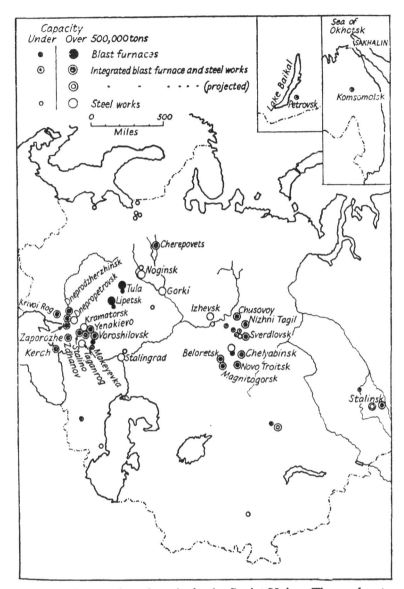

FIG. 11. Iron and steel works in the Soviet Union. The works at Kerch may not have been restored since the war. There are also a number of charcoal furnaces in the northern Ural Mountains

centre of iron-working. The Ural iron-ores occur in very numerous distinct deposits, some of them of quite small extent, over a distance of 600 miles from Ivdel' in the north southwards to Orsk. The operations of the Demidov family had been mainly in the central and northern Urals, but the ore-bodies are largest and richest in the south, and it is here that the most recent developments have taken place. The largest ore deposits are at Magnitogorsk, though the better ores of this deposit, enriched by the leaching away of the silica, are almost exhausted. In the future, reliance will be placed on the lower and poorer ores of this deposit as well as on other ore-bodies lying farther north in the Ural mountains. The effect of this is to call for more elaborate beneficiation plants and a more developed transport system.

As already noted, this region relies heavily on coking coal from the Kuzbas, which is not very high in quality, and also on supplies from Karaganda. It is claimed, however, that progress is being made in using the coal from the small Ural deposits to make a metallurgical coke. The seriousness of the fuel problem, which is in part at least a transport problem, is illustrated by the fact that the ancient charcoal-iron industry, which never died out completely, has been revived in the northern Urals, and new furnaces built to use charcoal fuel.

At present the chief integrated works are at Magnitogorsk, Chelyabinsk and Nizhni Tagil. A new plant is now being built at the southern end of the Urals at Novo Troitsk. In addition there are several smaller works, some of them dependent mainly on charcoal fuel, and a number of steel- and rolling-mills.

The third major centre of the iron and steel industry is the Kuznetsk basin, lying about 1,200 miles east of the Ural mountains, in western Siberia. It is based upon extensive coal deposits, but was developed with the object of using ores from Magnitogorsk. We have already seen that this has helped not only to deplete these ores, but also to produce acute transport problems. The policy of the Soviet planners has recently been to make this region independent of the Urals. A number of ore-bodies were discovered in the Kemerovo Oblast, within which part of the Kuznetsk basin lies. But these are small and some have been exhausted within a very few years. It is not yet clear that these deposits together contain enough ore to supply the existing furnaces for the rest of their working lives. Larger deposits have been found some 200 miles farther to the east, in the mountainous upper valley of the Yenisey river. The opening up of these ores has had to await the completion of a railway across the mountains from Stalinsk. This appears now to have been done, and a second large, integrated works is being built at Kemerovo in the Kuzbas in order to smelt these ores.[21] The Stalinsk works have recently been reported to be operating on 'local' ores.

At present the Stalinsk works, the second largest in the Soviet Union, is the only integrated plant in the Kuznetsk region, but, as we have seen, a second such plant is projected and may now be under construction. There are steel- and rolling-mills also at Novosibirsk.

In their efforts to exploit fuel and ore resources in the formerly undeveloped areas the Russians have established the smelting industry deep into Siberia, Kazakhstan and Central Asia. A blast-furnace and steelworks has been established at Temir-Tauz and Aktyubinsk in Kazakhstan, designed to use coal from the Karaganda field and ore from local deposits. Other works were established at Petrovsk, to the east of Lake Baikal, and yet another at Komsomolsk, in the Soviet Far East. Both use local materials which are known to be none too satisfactory. The exact status of these works is not clear at present. In the last few years an integrated works has been built at Rustavi, near Tbilisi in the Georgian S.S.R. It uses exclusively fuel and ore obtained in the Republics south of the Caucasus range. The broad policy behind these developments is to create regional bases of heavy industry and thus reduce the strain on railway transport.

European Russia, apart from the Ukraine, is noteworthy more for its steel- and rolling-mills than for its integrated works. The former are particularly numerous in the districts of Moscow, Leningrad and Tula, where their most important markets are to be found.

It is planned to increase yet farther the capacity of the iron and steel industries, and the question arises: where will the planners locate the new plant which they will be obliged to build? The past twenty years have shown them experimenting, as it were, with the fuel and ores of Central Asia, of eastern Siberia, of Transcaucasia and of other areas not hitherto developed, and a steel output of 56 million tons in 1957 indicates their success. Difficult problems, however, lie mainly in two directions: the raw materials—ore and coking coal—are abundant but mainly of qualities that cannot be used without extensive and costly preparation and beneficiation; and, secondly, they are widely separated both from one another and from the markets. In whatever geographical directions the industry is extended, the capital investment in plant and transport facilities will be disproportionately large.

It would appear, from the scanty evidence, that no great expansion may be looked for in the older centres, the Donbas and the southern Urals. The link between the Kuznetsk basin and the Ural mountains ore deposits appears now to be becoming tenuous. On the other hand, ore-bodies in the mountains to the east of Stalinsk are being opened up, and there are reports of extensive discoveries in Kazakhstan, and, with this ore in view, new plant is being built in the Kuznetsk region. Further, because of its relative nearness to China, a vast new development is projected in the Irkutsh region. Nevertheless much of the future expansion is likely to be obtained by modernizing and extending existing works rather than by erecting new.

Another area of current expansion is in the north-west of European Russia. The Pechora coal basin, though located mainly within the Arctic Circle, provides a coking coal. Pechora coal, and ore from several deposits in the neighbourhood of Murmansk, are now being used in the works newly built at Cherepovets, to the east of Leningrad. It may perhaps be suggested that a probable line of future development will be to bring together Pechora coal and the iron-ore of the northern Ural mountains. Such a development, however, presupposes a heavy capital investment in transport and in plant to prepare the materials for use. Current plans include the opening up of the Kursk deposits, to the south of Moscow, which have not hitherto been seriously worked, and the newly discovered deposits at Kustanay in north-western Kazakhstan. It is anticipated that the latter will supply about 9 per cent of the ore used in 1965. It is possible also that it will replace Magnitogorsk as a source of ore for the Kuzbas.

Soviet industry makes intensive use of its capital, its scarcest commodity. Labour is relatively abundant, and is used more lavishly and extravagantly than would be possible in Western Europe and North America. Russian statistics do not, like those of the United States, stress the productivity of labour. Instead, they emphasise the productivity of plant. And in this respect the Soviet record is a good one. An English observer wrote of the blast-furnaces at Magnitogorsk that

'the production of these furnaces is truly outstanding. The average production of the four furnaces of 26 ft. 9 in. hearth diameter . . . was 2,000 tons per day. . . . This rate of production is some 40 to 50 per cent higher than the production of furnaces of similar size in the U.K. and the U.S.A., operating on ores of similar iron content.'[22]

All this demonstrates the scarcity of capital and points up the difficulties which the Russians experience in using their materials, which are not only widely dispersed by are unusually intractable. This may to some extent explain and excuse the reversion, which occurred in 1955, from the Malenkov emphasis on the lighter industries to the older, Stalinist line which stressed the heavy industries.

CHINA

The early Chinese practice of metallurgy was marked by the development of the art of casting iron.[23] The casting of iron necessitates a temperature high enough to bring the metal to a fluid condition, and this, as we have seen, was not achieved in the West until the blast-furnace came into use in the later Middle Ages. Yet the earliest literary references carry iron-casting in China back to the sixth century B.C., and large castings survive from the tenth and eleventh centuries. There is reason to suppose that the Chinese owed their success, in part at least, to the practice, which they must have discovered empirically, of adding vivianite to the metal. This is a phosphatic mineral whose effect is to make the iron more fluid.[24]

Iron-working has survived in China on a craft basis until the present. The first installations of a modern iron and steel industry were established in the last decade of the nineteenth century. Little more was done until after the revolution which overthrew the Manchu dynasty and established the Republic. Several works, chiefly in Manchuria, were erected during the period of the First World War, but this was followed by the long period of civil war and Japanese invasion, when again little was built on the foundations that had been laid. In 1931, the Japanese overran Manchuria and in subsequent years penetrated deeply into China itself. Their war needs made it necessary to expand the existing industry and to extend the number of iron and steel works.

The resources of China resemble in certain respects those of the Soviet Union. There are a few deposits both of ore and of coking coal that are large enough to justify the building of large-scale works in their vicinity, but there are also many small deposits scattered through the rest of the country. The Japanese were the first people to make a really extensive use of other smelting devices than the blast-furnace. They built a large number of Krupp-Renn furnaces (*see* above, page 64), which operate on a small scale and are tolerant of poor fuel. They are extravagant in their demands on labour, but in China this factor was of much less importance even than in the Soviet Union.

The defeat of the Japanese has been followed by the coming to power of the Communists. In China, as in other countries under Communist rule, heavy industry has been intensified; the existing iron and steel works have been extended and new ones founded. The production of pig-iron has risen from about 780,000 tons in 1938, the most successful pre-war year, to 27,500,000 tons in 1957. Steel production rose within the same period from about 500,000 tons to 18,450,000. China is unique in having a pig-iron output considerably larger than steel output. This implies, on the one hand, a shortage of steel scrap, and, on the other, a relatively large demand for cast-iron goods. These two facts together denote a relatively primitive economy.

The most abundant ores are at Anshan and Kung-ch'ang-ling in southern Manchuria. The best of these ores are good-quality magnetite and haematite, but, like the corresponding ores near Lake Superior, they are nearing exhaustion, and the remainder of these very extensive deposits is made up of low-grade, highly siliceous ores. It is, nevertheless, on these south Manchurian ores that much of the Chinese industry is based.

The iron ore deposits in Inner Mongolia are rather small in reserves and too remotely located to be of value in the near future. There are also small deposits in Hopeh, but, apart from the Manchurian ores, most of the reserves are in the Yangtze valley. Most of these are fairly close to the river, and are thus amongst the more accessible ores of China, but, with the exception of the Ta-yeh ores, near Hankow, the ore-bodies are individually very small. Not even the Ta-yeh deposit is large enough to support a modern works. It was in view of the large number of widely scattered, small deposits, and of the super-abundance of cheap labour, that the Krupp-Renn and similar processes have become important.

China is proverbially rich in coal, but the proportion of her coal deposits that make a satisfactory coke is not high. The rich coalfield at Fushan in southern Manchuria does not yield coking coal. As in the Soviet Union, any large-scale expansion of the iron and steel industry must be preceded by the extension of the network of transport and communications. A recent writer noted that at Anshan, the largest of the Chinese iron and steel works, the 'transportation requirement is higher than at the Russian plant at Kuznetsk.'[25]

The iron and steel works in China today fall into one of three groups, the Manchurian, the North Chinese and those of the Yangtze valley. In the first of these the most important centre, and the largest in the whole country, is at Anshan.[26] These works were founded by the Japanese in 1917, but did not prosper or expand greatly until they again fell into Japanese hands at the time of the Manchurian crisis. They were partially dismantled by the Russians in 1945, but re-equipped and restored by them after China had come within the Communist sphere. As we have already seen there are large reserves of rather poor ore in the vicinity but coke has to be brought large distances. Nevertheless, the capacity of the works is said to have been expanded to about 1,500,000 tons of pig-iron a year. Steel production is smaller. The Penchihu works, also in south Manchuria, are very much smaller than those of Anshan, but are very much more favourably placed with regard to raw materials.

The second group of works lies in the North Chinese provinces

F

of Hopeh and Shansi. There are numerous small ore deposits in this region as well as extensive coal deposits in Shansi. The oldest plant lay at Shihchingshan, a short distance to the west of Peiping. There is a small iron and steel works at Taiyuan, in Shansi, and a steel works at Tientsin, between Peiping and the coast. The Japanese established other works in North China, including a number of Krupp-Renn furnaces, but it is uncertain whether these are still active. A large, integrated works has been begun at Paotou, in Inner Mongolia, but it is said to be experiencing great difficulty in its supply of coking coal.[27]

The last group of works lies in the valley of the Yangtze, a part of China that has always been the most open to Western influences. As in northern China, there are numerous deposits, most of them small, of iron-ore. Coking coal can be obtained locally, and the river itself provides the cheapest and most convenient means of transport. Several works were built at Shanghai, in Anhwei and in Hupeh. The earliest modern works to be established in China was built at Hankow, in Hupeh province, where there is a well-placed deposit of good-quality ore at Ta-yeh. A large integrated works is under construction near Tangtu Hsien, on the Yangtze, and another has been begun at Hankow.[28]

Political and military events during the Second World War emphasized the importance of the far western provinces of China Proper. Iron and steel works, intended presumably for the supply of military equipment, were built near Chungking and Kunming. Whatever may have been their post-war vicissitudes, these works have been modernized and extended and are now again in production.

One is inclined to say that the extension of the modern steel industry in China faces such difficulties that it might be judged impossible. But this is in the context of Western competitive industry. Chinese industry is neither competitive nor completely Western. The severe regimentation and the low levels of remuneration of a vast population, amid which hidden unemployment has been normal, alters the situation. The declared objective of the Peiping government was to produce over 10 million tons of steel in 1958. Such a rate of expansion would be impossible by means of modern methods, because the construction of plant alone takes a long period of time.

China has reverted to what is, in effect, a craft industry. The Peiping radio has announced that 'in the hilly areas 3,000 people will be organized in search of mineral deposits. . . .' In Yunnan, South-west China, more than 10,000 small blast-furnaces and local-type iron-smelting furnaces were built up to the end of August (1958).[29] Similar reports are available from other parts of China. Small steel-making plants are being established in great numbers.

The era of the backyard furnace is apparently over, and iron and steel expansion now follows more orthodox lines. Perhaps the most that can be said against the Chinese method of expanding output is that it made such excessive demands on labour as to be completely impracticable outside China. It most definitely allowed the fullest use to be made of the small and scattered deposits of ore and fuel.

The Communist sphere includes also North Korea and Northern Viet Nam. North Korea is potentially quite important. It has a number of ore deposits, of which the Musan reserves are of fairly low grade but very large.[30] These and the numerous ores on the west coast of Korea were formely mined and shipped to Japan. In 1944 the total ore production reached nearly 3·5 million tons. A number of small blast-furnace works was established here. The Kyomip'o works, at Kenjiho, on the west coast, was equipped with steel-making- and rolling-mills. The others were blast-furnace works, which smelted the Musan ore and formerly exported pig-iron to Japan. During the Second World War the Japanese built an immense number of small, low shaft-furnaces in northern Korea as well as some Krupp-Renn plants. It does not appear that any of these are in production today.

Northern Viet Nam is much less well endowed than North Korea. There are numerous ore deposits, most of them small. Though the ore has been used by native craftsmen, no attempt seems to have been made to exploit it commercially.

[1] Albania is omitted, as it has no iron and steel industry.
[2] Nicolas Spulber, *The Economics of Communist Eastern Europe*, New York, 1957; Jan H. Wszelaki, *Fuel and Power in Captive Europe*, Mid-European Studies Centre, New York, 1952.
[3] See *Studia z Dziejów Górnictwa i Hutnictwa*, Tom I, Polska Akademia Nauk, 1957.
[4] P. H. Seraphim, *Industrie-Kombinat Oberschlesien*, Koln-Braunsfeld, 1953; *Ibid.*, *Deutschlands verlorene Montanwirtschaft*, Stuttgart,

1955; J. M. Montias, 'The Polish Iron and Steel Industry', *The American Slavic and East European Review*, XVI, 1937, 300–22. N. J. G. Pounds, *The Upper Silesian Industrial Region*, Slavic and East European Series, XI, Indiana University, 1959.

[5] M. Gardner Clark, *Report on the Nowa Huta Iron and Steel Plant named after Lenin, near Cracow, Poland*, privately printed and circulated, Cornell University, 1957; N. J. G. Pounds, 'Nowa Huta', *Geography*, XLIII, 1958, 54–6.

[6] O. Quadrat, 'La situation de l'industrie siderurgique en Tchecoslovaquie', *Revue de Metallurgie, Memoires*, XXXII, 1935, 469–87; C. Berthelot, 'La Roumanie et la Tchecoslovaquie minières et metallurgiques', *Ibid.*, 156–70, 200–9; *The Iron and Steel Industry of Czechoslovakia*, Prague, 1930.

[7] Ivan Avsenek, *Yugoslav Metallurgical Industry*, Mid-European Studies Centre, New York, 1955; *Iron and Steel in Yugoslavia 1939–1953*, *loc. cit.*, 1953; *International Coal Trade*, U.S. Department of the Interior, XXVII, 1958, No. 6, 26–27.

[8] 'Steel Developments in Rumania', *Monthly Statistical Bulletin*, B.I.S.F.

[9] 'Steel Developments in Eastern Germany', *Monthly Statistical Bulletin*, B.I.S.F., October, 1954.

[10] R. Portal, *L'Oural au XVIII siècle*, Collection historique de l'Institut d'Études Slaves, Paris, 1950.

[11] *Economic History Review*, IV, 1951, 252–5. See also Ludwig Beck, *op. cit.*, III, 1122–57; James Mavor, *An Economic History of Russia* (London), 1914, I, 434–7.

[12] Eli F. Heckscher, 'Un grand chapitre de l'histoire du fer: le monopole suedois', *Annales d'Histoire economique et sociale*, IV, 1932, 127–39; 225–41.

[13] Neumark, 'Die Russische Kohlen-und Roheisen-Industrie mit besonderer Berucksichligung der sudrussichen Verhallnisse', *Stahl und Eisen*, XXI, 1907, 62–68; 110–22.

[14] For a general survey of economic growth in the Soviet Union, *see* Harry Schwartz, *Russia's Soviet Economy*, New York, 1954.

[15] This follows M. Gardner Clark, *The Economics of Soviet Steel*, Harvard University Press, 1956; for both the history and economics of this project, see Franklyn D. Holtzman, 'Soviet Ural-Kuznetsk Combine: a Study in Investment Criteria and Industrialization Policies', *The Quarterly Journal of Economics*, LXXI, 1957, 368–405.

[16] *See also* Demitri B. Shimkin, *Minerals, A Key to Soviet Power*, Harvard University Press, 1953, 174–94.

[17] M. Gardner Clark, *op. cit.*, 178.

[18] The Russian Steel Industry, *Steel Review*, April, 1957, 24–32; *The Russian Iron and Steel Industry*, The Iron and Steel Institute, Special Report No. 57, London, 1956; *Steel in the Soviet Union*, American Iron and Steel Institute, New York, 1958. *See also Mineral Trade Notes*, U.S. Bureau of Mines, Sept. 1957.

[19] V. Holubnychy, 'The Present State of the Ferrous Metal Industry', *Ukrainian Review*, 1957, No. 4, 5-14.

[20] T. P. Colclough, 'The Urals' Steel Plants', *Steel Review*, April, 1957, 40-48.

[21] Figures quoted by M. Gardner Clark, *op. cit.*, 182-3, suggests that this ore is too limited in extent to justify such an expansion. It is likely, however, that recent prospecting has led to an increase in estimates of its size.

[22] T. P. Colclough, *op. cit.*, 47-8.

[23] W. Gowland, in *Archaeologia*, LVI, 1899, II, 267-322.

[24] Thomas T. Read, 'The Early Casting of Iron', *Geographical Review*, XXIV, 1934, 544-54.

[25] Muzaffer ErSelçuk, 'The Iron and Steel Industry in China', *Economic Geography*, XXXII, 1956.

[26] Thomas T. Read, 'Economic-Geographic Aspects of China's Iron Industry', *Geographical Review*, XXXIII, 1943, 42-55.

[27] T. Shabad, *China's Changing Map*, New York, 1956, 59-62.

[28] *Survey of World Iron Ore Resources*, United Nations, 1955, 326-8.

[29] From text of monitored Chinese radio, U.S. Department of State, Sept. 1958; see also *New York Times*, Nov. 15, 1958.

[30] *Ibid.*, 302-3.

8

THE UNDERDEVELOPED COUNTRIES

WE HAVE seen already that the production of iron and steel goods is distributed very unevenly over the land surface of the earth. This is reflected in the distribution of consumption of such goods. The contrast between the United States and the Far East is extreme. On average, the American consumes per year at least twelve times as much steel as the average inhabitant of the Far East. The scale of consumption in the Middle East, Africa and Latin America is not greatly superior to that in Asia. The following table shows the apparent consumption per head of the population of steel goods in 1948:

	Kilogrammes
U.S.A.	518
Europe	110·5
U.S.S.R.	86·3
Latin America	24·1
Africa	12·1
Middle East	11·3
Far East	4·6

(From *European Steel Trends*, Geneva, 1949, 130–6.)

It is certain that in the ten years that have since elapsed the distance between the biggest and the smallest consumers has lengthened.

Yet in all these regions the trend in the consumption of iron and steel goods has been upward. It was interrupted by the Second World War, but a recovery in effective demand was manifest afterwards. This upward trend was, of course, most marked in the U.S.A. and Europe, but was apparent also in the Middle and Far East. Recent heavy capital investments in Africa, the Middle East and

India have increased yet more the consumption of iron and steel in these areas.

Since the end of the Second World War, the older centres of iron and steel production have expanded their capacity. Fears were expressed in 1949[1] that the rate of demand in the rest of the world for European steel goods might slacken, and that the European industry might find itself over-equipped. Fortunately, the ceiling in demand which was anticipated by the Economic Commission for Europe proved to be illusory. For this the Korean War and the deterioration of the international situation were in part responsible, but the increasing capital investment by the leading industrial countries in the territory of the less developed, however mixed may be their motives, certainly demands increasing quantities of steel. Subsequent reports[2] revised the rather gloomy prognostications of 1949, and their tone of qualified optimism has been answered by further plans for expansion.

But plans for extending the iron and steel industry are not confined to those countries which have long been its centres. There is a close relation between the production of steel and the standard of living. The creation of an iron-smelting and steel-making industry does not in itself bring about a rise in living standards. Much depends on control over the industry and the uses to which the steel is put. It is nevertheless true that the steel industry is basic to many other industries, including the engineering and the chemical, and also to the whole organization of transport. Expansion of engineering and chemical industries may result in more and better agricultural equipment and fertilizers. It is certain that the latter will not be expanded without a previous growth in steel production.

The Communists are correct in stating that the growth of steel production is basic to expansion of the light and consumer goods industries. It is, unfortunately, also a necessary part of plans for military preparedness. It is probable that military considerations have been a factor in most recent plans for expanding steel capacity, and not least in those of the underdeveloped countries. It scarcely needs to be said that military uses constitute a sterile employment of steel and one which, though figuring in national income calculations, does not contribute to the level of welfare.

It has been pointed out earlier (page 62) that a steel industry has a prestige value quite apart from its military potential. And no underdeveloped country that has developed an iron and steel industry has failed to derive the maximum publicity from it.

Great difficulties face any underdeveloped country that seeks to create or to expand an iron and steel industry. First of these is the supply of raw materials. By a strange quirk of nature, it is these underdeveloped countries that are best endowed with iron-ore. A recent United Nations estimate attributes to the arbitrarily defined underdeveloped countries considerably over half the world's iron-ore. The following table illustrates this contrast between natural endowment and actual industrial development:

	Percentage of	
	Probable iron-ore reserves	Crude steel production
Africa	13·6	0·4
Asia (excl. Japan)	22·9	0·9
Latin America	18·7	0·6
Total underdeveloped areas	57·3	4·6
Europe*	23·3	27·8
North America	10·6	54·8
U.S.S.R.	8·4	10·9
Australasia	0·4	0·8
Total industrialized areas..	42·7	95·4

(From *World Iron Ore Resources and their Utilization*, United Nations, 1950, 17.)

* Excludes the Iberian Peninsula and certain East European countries. These are included in the total for 'underdeveloped areas'.

This relative abundance of ore is not matched by comparable resources of fuel. Indeed the pattern is almost reversed. The industrialized countries have large reserves of coal, in which, of course, we see one reason for their industrial growth, with a high proportion of it of coking quality (*see* table opposite). Thus we have some 60 per cent of the iron-ore in the underdeveloped countries, but only 9 per cent of the world's coal, and most of this is of a quality unsuitable for making metallurgical coke.

The second problem, which cannot be enlarged upon in this book, is the supply of capital. During the Industrial Revolution the countries of Western Europe succeeded in building up the iron and other industries on the basis of their internal resources in materials, capital and labour. Their industrial growth was necessarily slow, but in the course of between two and three centuries they have evolved machines of such complexity as, for instance, the continuous strip mill. Any underdeveloped country has the means to build and operate a primitive blast-furnace and refinery. They could not, on the basis of their own human and material resources, erect a

	Probable iron reserves (in millions of tons of iron content)	Probable* coal reserves (in millions of tons)	Proportion of coal of coking quality
Europe	5,562	533,140	High
North America ..	2,640	2,093,055	High
U.S.S.R.	2,027	1,443,000	High
Australasia	130	15,298	High
TOTAL INDUSTRIAL-IZED AREAS	10,359	4,084,493	—
Africa	3,609	74,765	Moderate
Asia	6,988	332,308	Low
Latin America ..	5,763	37,166	Low
TOTAL UNDERDEVEL-OPED AREAS	16,360	444,239	—

(From *World Iron Ore Resources and their Utilization*, 66–7.)

* The potential ores are many times larger.

rolling-mill. Unless the underdeveloped countries take up the iron industry at a technical level that has long since been abandoned in the West—a course that is in the highest degree improbable—they are obliged to obtain from the industrialized countries not only plant and equipment, but also the skills with which to use them. These may be paid for by current exports or by loans or they may be requited in some less direct manner by fulfilling political services. The provision of such skills and equipment becomes a political weapon, which appears to be used more and more by both sides in the Cold War, in their efforts to secure the services and support of the uncommitted third. In this way the Soviet Union has equipped the Polish Nowa

Huta works and contributed materially to the establishment and operation of plant in China, India, and in the countries of Eastern Europe. On the other hand, the investments of the United States, though not politically motivated to the same extent as the Soviet, are doubtless expected to restrain the recipients from an over-zealous association with the Soviet Union. Some recent developments have been financed by the International Bank. Few countries are capable of providing the mechanical plant needed by the modern iron and steel works. The United States is the biggest producer and probably also the biggest exporter of such equipment. The Soviet Union, certainly less skilled in this respect than the U.S.A., has exported equipment to her Communist allies. Western Germany has in recent years become an important supplier of steel-mill equipment, with large orders from India.

A final problem is that of the scale of the planned undertaking. The economies to be derived from large-scale production are more conspicuous in the field of iron and steel production than in that of most other branches of industry. The branch of steel production that contributes the most directly to improved living standards is thin and very thin strip, such as is used for containers, for automobile bodies and for most domestic goods.[3] The economies of continuous rolling are immense, and the huge consumption of these goods in the U.S.A. has resulted as much from the diminished cost of production as from a rise in personal incomes. But, as we have seen (page 66) the continuous strip is large and very productive. It presupposes a large steel-mill and also a large market. Few of the underdeveloped countries could absorb the output of a single integrated iron and steel works with continuous strip mill. None of them could produce at a price anywhere near the European or American the whole range of rolled-steel goods that modern civilization demands. Only with an immense improvement in living standards could most of the countries which are to be discussed in this chapter consume annually enough steel for them individually to be self-sufficient at an economic price.

There are, of course, two qualifications. Politics and prestige may together bring about developments that on simple economic ground would appear impracticable. Secondly, the technological conditions of the modern integrated works are assumed. It is difficult to visualize a means of rolling steel that could be operated as cheaply on a smaller scale than the modern mill. But there are, of course, small-scale alternatives to the blast-furnace (page 64), and not all these are extravagant or difficult to operate. The various sponge-iron processes have recently been held out as a means of meeting the problems not only of scale of production, but also of low-grade fuel in the underdeveloped countries.[4]

In the remainder of this chapter the future is discussed of the iron and steel industry in the underdeveloped countries. In the previous chapter the countries of the Soviet Sphere were discussed. These included not only China, North Korea and Northern Viet Nam, which are clearly underdeveloped, but also the Communist countries of Eastern and South-eastern Europe, some of which should also belong to this category. On the other hand, this chapter does include a consideration of Australasia and Japan, which, in the United Nations statistics already used, are classed among the industrialized countries. Our definition of the underdeveloped countries is arbitrary in the extreme. To make them synonymous with what is left over, after North America, Europe and the Soviet Sphere have been studied, makes too great a concession to convenience of organization. But within a small book there seems to be no alternative.

As defined, then, the underdeveloped countries are divided into four groups: Latin America; Africa; Middle East, India and South-East Asia; and Australasia. Within each of these four groups, problems and difficulties have much in common.

LATIN AMERICA

Latin America is, as we have seen (page 169) relatively rich in reserves of iron ore.[5] Among these are the abundant, high-grade ores of Minas Gerais, and the haematite of Chile and Venezuela, which are good enough for export to the United States. On the other hand, the South American continent is the least endowed with coal. Some of the coal of Chile and Colombia is of coking quality; Brazilian coal is poor, and in the rest of South America it is doubtful whether coking coal exists. In the whole of Latin America, only Mexico has good-quality hard fuel available in quantity.

The following table gives the population and the consumption of steel goods of some of the Latin-American countries for 1950–1.

			Population (in thousands)	Steel consumption (in thousands of metric tons)	Consumption per head (in kgms.)
Argentina	..	..	17,641	1,111·7	63·0
Brazil	..	..	52,124	874·9	16·8
Chile	..	..	5,920	206·2	34·8
Colombia	..	..	11,260	152·2	13·5
Mexico ..	..	..	26,332	737	28·0

(Based on *A Study of the Iron and Steel Industry of Latin America*, Vol. I, United Nations (1954), pages 84–6. Figures relate to 1950 or 1951.)

The low level of consumption is a major obstacle in the way of developing industry. This can be expected to rise slowly, but it will necessarily be long before most of the Latin-American countries could, on the basis of market considerations alone, establish an integrated iron and steel plant. On the other hand, an over-reliance on imports has its disadvantages. During the war years and for much of the post-war period the market for steel was extremely tight, and the normal needs of the Latin-American countries became difficult to satisfy. This was a major factor in leading certain Latin-American countries to satisfy at least part of their needs from domestic sources. Hitherto, only four Latin-American countries have taken steps to establish iron and steel industries, and of these only Brazil and Mexico have made any considerable progress. The expansion of output in recent years is shown in the following table:

(*thousands of metric tons*)

	1938 Pig-iron	1938 Steel	1945 Pig-iron	1945 Steel	1955 Pig-iron	1955 Steel	1957 Pig-iron	1957 Steel	1965 Pig-iron	1965 Steel
Brazil	122·4	92·4	259.9	205·9	1,198	1,380	1,320	1,760	2,400	2,896
Mexico	98·1	88·6	194·0	191·4	356	838	485	1,136	942	2,403
Argentina	—	10·0	18·0	15·0	40	240	37	400	663	1,360
Chile	—	—	13·4	27·0	282	320	420	440	309	441
Colombia	—	—	—	—	109	85	139	130	242	204
Venezuela	—	—	—	—	—	—	—	—	334	625
Peru	—	—	—	—	—	—	—	—	20	81

This rate of production is small enough, but the rate of increase in production surely suggests that the iron and steel industries have a future in Latin America.

Brazil: It is in Brazil that the most ambitious programmes have been prepared for developing the iron and steel industries.[6] Brazil has the largest population of any Latin-American country as well as the largest consumption of steel goods. She has furthermore the most extensive deposits of high-grade ore in the continent, and also a generous share of Latin America's scanty coal resources. Before the Second World War there were several small iron and steel undertakings, with a collective output of under 100,000 tons of steel. They were located near the Itabira ore deposits, and some of the works used charcoal as fuel.

In 1940 the Companhia Siderurgica Nacional was formed with the object of building a large modern integrated works. A site was chosen at Volta Redonda, a short distance inland from Rio de Janeiro and on the railway from Rio to São Paulo. Ore is obtained by rail from Itabira, and fuel in part from the Santa Catarina field of southern Brazil, in part from the U.S.A. The works lie close to their chief market in south-eastern Brazil. They were in production by 1946, but have since been greatly extended. The ultimate objective is said to be a production of ten million tons of steel yearly. Some of the older works in the neighbourhood of Itabira have been extended, and new steel works have been built.

Mexico: Like Brazil, Mexico is well endowed with ore, and her reserves of coking coal are better than those of any other Latin-American country. The supply of metallurgical coke is, however, said to have been inadequate, and the two integrated iron and steel works have been obliged to work considerably below capacity. A major difficulty in the expansion of the iron and steel industry is the poor transport facilities between the sources of materials and the blast-furnace works.

Argentina: Relative to its population, Argentina has been one of the heaviest consumers of steel in Latin America, and this factor, combined with the strength of national feelings, has led Argentina to develop an iron and steel industry. The potential market may well be large, but the resources are quite inadequate. The few coal-fields along the foot of the Andes and the several scattered ore-bodies are small and low-grade. A small blast-furnace works in the north-western province of Jujuy uses charcoal fuel, but the new industry that is developing in the east of the country is overwhelmingly dependent on imported coal and iron-ore. It was presumably to facilitate the use of imported materials that a new, integrated works is being built on the banks of the navigable Parana river at Villa Constitucion. A second integrated works has also been begun a few miles downstream at San Nicolas.

Chile: The natural resources of Chile to support an iron and steel industry are very much better than those of Argentina. Large reserves of ore are found a short distance inland in northern Chile, and have been mined for export to the U.S.A. for some forty years. Coal, partly of coking quality, is found in the coastal region of southern Chile. A small blast-furnace industry was established near Valdivia in 1910, but the only integrated works was established at Huachipato since the end of the Second World War. The works lie close to the coast, at no great distance from the Cape Lebu coal-field. The coal from this source is of poor quality but can be used if supplemented by imported fuel. The ore is shipped from the mines of northern Chile.

Lack of resources and the high cost of overland transport have not deterred other South American countries from formulating plans. The most recent to enter the field is Colombia.[7] A small, integrated works has recently been built at Belencito. It lies close to deposits of *minette*-type ore, which are said to be extensive, but is dependent to some degree on imported ore. Peru is building an iron and steel works at Chimbote, on the coast of northern Peru, with low-shaft electric furnaces, and Venezuela has plans for a similar works at the junction of the Orinoco with the Caroni rivers. This project has not yet been begun.

When Colombia applied to the International Bank for assistance in building its Belencito works, its representatives were told that the project was uneconomic. The same could probably be justly said of other works, existing and proposed. Indeed, it may be said that only Mexico, Venezuela, Brazil and Chile appear to be in a position to make steel at a price no higher than that of imported steel. 'If Latin-American countries could establish efficiently operated steel industries serving regional markets, rather than domestic markets alone,' a United Nations publication reported, 'the potentialities for steel production would be expanded in Brazil, Venezuela and Chile particularly. . . . If, on the other hand, Chile, Argentina, Brazil, Peru, Venezuela and Colombia should build steel plants, the markets for each would be so limited as to reduce the efficiency of production in Latin America as a whole.'[8] Unfortunately nationalism is a powerful force in Latin America; one cannot foresee Argentina, Peru and Colombia willingly putting such a weapon into the hands of their neighbours and rivals.

AFRICA

The similarities between the geological history of Africa and of South America are reflected in their mineral endowment. Both are relatively rich in iron-ore and poor in coal. Geological surveying has made too little progress for any list of deposits to be final and complete, but it appears at present that the greater part of the deposits of good-quality ore are to be found in Morocco and Algeria, in tropical West Africa and in the Union of South Africa. Deposits which are known to exist in Central Africa may prove, when more carefully prospected, to have reserves as large as these.[9]

The Algerian ores (see pages 46–7) have been an important factor in the rise of the smelting industry of Western Europe, and they continue to be exported in considerable quantities—1,028,000 tons of metal content in 1956. Morocco and Tunisia have exported ore in smaller quantities. It never was a part of French plans to use these North African ores as a basis for a local smelting industry.

The export of ore from the extensive deposits of several West African territories is at present increasing, but here also no attempt has been made, or is likely to be made, to smelt them locally. There is no satisfactory metallurgical fuel in these areas, and the market, a thinly scattered population with a low purchasing power, would certainly not justify such an undertaking in the near future.

Union of South Africa: It is only in the Union of South Africa that a modern industry has been established.[10] The natural endowment of raw materials appears to be more propitious here than elsewhere in Africa and the *per capita* consumption of steel is the highest by far in the continent. This is due in part to the more strongly developed transport net, the rise of light industries, and above all to the growth of mining. The gold-mining industry has become a major consumer of steel, and exercised a strong influence on the location of new plant.

The iron and steel industry of the Union of South Africa is today carried on at three places: at Newcastle in Natal, and at Pretoria and Vereeniging in the Transvaal. All three are well placed to use domestic ore and fuel, and are close to the important market formed by the mining industries of the Veld. The Transvaal industry is by far the largest. It was founded before the First World War and was greatly expanded during the 1930's when the Pretoria works were established. They use ore from nearby mines supplemented by a very high-grade ore, resembling that of Mesabi, which is brought from Thabazimi, to the north-west of Pretoria. The supply of coking coal presents greater difficulties. The Witbank field, in eastern Transvaal, is easily worked, but does not yield a good coking coal. The Newcastle coal-field of Natal produces a good-quality fuel, but is more costly to mine and very much more distant from Pretoria. Coal from both these sources is blended at the coke-ovens at Pretoria.

The focus of South African iron-smelting and steel-making is moving southward to the valley of the Vaal river. A steel-and rolling-mill was established here at Vereeniging early in the present century. At first it used scrap produced mainly by the local mines and engineering works. Later this was supplemented by pig-iron brought in by rail from Natal. The smelting of ferro-manganese, using the Postmasburg ores, was then established, and now a fairly large integrated iron and steel works has come into production, at Vanderbijl Park, to the west of Vereeniging, with a capacity of 730,000 tons of steel.

The third important centre of smelting and steel-making is Newcastle in Natal. It has ore and good coking coal nearby, and the cost of assembling the materials is the lowest in the Union. A blast-furnace works, established here soon after the First World War, has recently been modernized. It is a relatively low-cost centre of production, and the works have limited themselves to smelting, and ship most of their pig-iron to the steel works at Vereeniging.

The Union of South Africa is unusual in having no works located on the coast. Unlike the other 'new' countries, South Africa has her markets mainly in the interior, where, fortunately, the raw materials are available. The high cost of rail freight virtually excludes the coastal cities of the Union from the market for Transvaal steel, and European steel has here a slight price advantage.

Southern Rhodesia: The Rhodesias lie at a considerable distance from the coast and, like Transvaal, have both the natural resources of steel production and a market for steel goods. A very small steel works has recently been established at Que Que, between Bulawayo and Salisbury. It uses coal from the Wankie field and ore from nearby deposits. Total demand at present could justify only a small and, presumably, an expensive plant. On the other hand, the high transport charges on goods imported through the Portuguese ports make imported steel goods extremely expensive.

Despite the success of the South African industry and the experiment in Rhodesia, it would not seem likely that other modern iron and steel plants will be located in Africa. The market is likely to remain so small that it would not be practicable, except in the Union of South Africa, to build even a medium-sized integrated works. On the other hand, imported steel is likely to be extremely expensive in all except the coastal regions of Africa, and the temptation will remain to develop an iron and steel industry wherever the materials are present.

ASIA

Some consideration has already been given in the previous chapter to the underdeveloped countries of Asia. It remains here to examine the prospects for industrial development of the Middle East, South and East Asia and, somewhat illogically, Japan.

Middle East: The Middle East is poorly endowed with iron-ore (*see* page 41) and coal resources are negligible. On the other hand, petroleum and natural gas form excellent fuel for steel-making, and the brash, young nations of this region would certainly derive pleasure if not also profit from building up an iron and steel industry. Apart from Turkey (*see* page 114), only Egypt has taken steps in this direction. A small integrated, works was begun in 1954 at

Helwan, on the Nile above Cairo, and came into production about four years later. Much of the plant is being supplied by West Germany. Ore is expected to come from desert deposits near Aswan and fuel from Western Europe. The plant was described by Egyptian authorities as of 'about the smallest economic size', with two blast-furnaces with a capacity of about 265,000 tons of pig-iron, and a steel works, operating mainly on the basic Bessemer process and producing about 220,000 tons. 'There is,' reported *The Times*, 'a surprising sense of achievement about the busy complex of installations which has sprouted on the firm desert floor near the riverside village of El Tibin',[11] and this sense of achievement may prove to be the most important product of these works.

India: The prospects for the iron and steel industry in India are incomparably better than in any other among the underdeveloped countries. Reserves of ore are large and of good quality (page 44). These are distributed over the Deccan plateau, but the largest reserves and the biggest production are in the Singhbhum-Mayurbhanj ore-fields of the Chota Nagpur region.[12] On the other hand, coal reserves—especially of coking coal—are not proportionate. The most abundant resources are in the Jharia coal-field, which lies about a hundred miles to the north of the Chota Nagpur iron-ore belt.[13] Unless inferior qualities of coal can be used, the future expansion of the Indian iron and steel industry may be seriously restricted by a shortage of fuel.

The modern Indian ore and steel industry was founded by Jamshetji Tata. His works at Jamshedpur, about midway between the Singhbhum ore and the Jharia coal, were begun in 1911, and have since remained the largest unit in India. The cost of assembling raw materials is low and labour is relatively cheap. Steel output in recent years has been about one million tons, but the plant is currently being extended and its capacity increased to two millions.[14]

A second centre of iron and steel production is on the Raniganj coal-field in the Damodar valley. A blast-furnace works was built here long before the end of the nineteenth century, but the area became important only with the construction of the Burnpur integrated works between the two World Wars.

Lastly, a very small smelting and steel-making works has been built at Bhadravati, in western Mysore. Charcoal is used successfully to smelt a high-grade ore. There are no coal resources here, and shortage of timber is likely to restrict expansion.

The steel production of India in 1957 was about 1,916,000 tons a year. Steel consumption has always been low, so that the supply of scrap is much smaller than in countries with more developed industry. This necessitates a relatively large smelting industry, and thus increases the pressure on the scanty reserves of coking coal. Despite this difficulty, output was expanded to 5,971,000 tons in 1963, though plans called for even more than this. Indian industry has the great advantages: good ore, a large market, the economies of large-scale production and the knowledge that it has already achieved technical and financial success with a minimum of outside help. Three large integrated works are now coming into production at Rourkela, Bhilai and Durgapur.

South-East Asia: In most of South-East Asia iron-ore resources are small, and many of them are either difficult of access or of poor quality. Only the Philippine Islands form an exception. Fuel suitable for blast-furnace use is rare throughout this region. Most of the countries, even in their wildest flights of optimism, have not proposed the establishment of an iron and steel industry—except Indonesia. It has recently been reported that work on an iron and steel plant was to begin in 1958 with capital supplied by the Soviet Union and Western Germany.[16] Even within this optimistic young republic warnings have been given that there is no suitable coal and that much of the ore is intractable. Of all countries that in recent years have planned to build an iron and steel industry, Indonesia would seem to have the smallest chances of success. Not only does she lack the materials; her recent acts are well calculated to repel foreign technicians and discourage foreign capital. In Thailand a small charcoal blast-furnace is in production near Tha Luang, and supplies iron for a very small open-hearth and rolling plant.[17]

Japan: Japan is discussed here because of the convenience of organization, not because Japan is an underdeveloped country. Her *per capita* national income is low, but in other respects Japan must be considered among the industrialized nations.

The first Japanese iron and steel works was completed at Yawata in 1901.[18] It was a government undertaking, built primarily to satisfy military needs. Although a number of privately owned steel works was established during the following years, the industry nevertheless grew slowly. Indeed, until 1932 the Japanese smelting industry remained inferior in size to India's. The varied industrial development of Japan, particularly ship-building, combined with the nation's military pretensions to increase greatly the demand for steel, and the steel-making industry grew more rapidly.

The natural resources of Japan for heavy industry are, however, inferior to those of India. The ore reserves are restricted in total amount and consist of small and widely scattered deposits. A large proportion of these deposits are made up of iron-sands, associated with former beach-lines and found now near the coast. At the present time these sands are being intensively exploited. It is said that the total amount of prospected ore in Japan amounts to only 25 million tons, with 11 million tons of iron content.[19] If all allowances are made for the incompleteness of the figures available, it is apparent that Japanese resources, if used exclusively, could suffice for only a year or two.

Nor are fuel resources in any way superior. The coal is mainly Tertiary in age, and contains a high proportion of ash and of volatile matter. Over a third of the coking coal used in Japanese iron works is imported from the United States.

Among the major iron- and steel-producing nations Japan was formerly unique in her heavy dependence upon imported ore, pig-iron and scrap. Nowhere else was the ratio of pig-iron to steel production as low as in Japan. This is, of course, explained not only by her lack of ore and the economies of importing scrap-metal that has already been refined, but also of the need to economize in coking coal. The table on p. 182 illustrates this situation.

Pig-iron was imported chiefly from Manchuria, China and Korea, and a very large proportion of the scrap came from the United States. Here the Federal Government in 1940 prohibited further scrap exports to Japan, and a result of this was the accelerated production of pig- and sponge-iron in Krupp-Renn, low-shaft and other inexpensive and quickly built smelting devices in the territories controlled by Japan, as well as a more intensive exploitation of the very small domestic ore reserves.

Enough has been written to show how slender and precarious

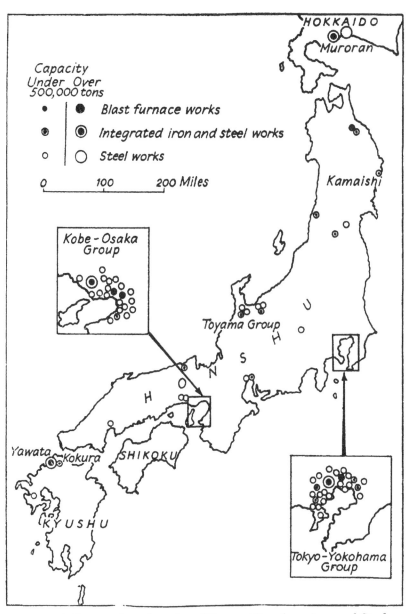

Fig. 12. Iron and steel works of Japan. (Based on *Iron and Steel Works of the World*, ed. H. G. Cordero, and *Statistical Year Book for 1957*; The Japan Iron and Steel Federation)

was the resource base on which Japan built up her iron and steel industry. After her defeat in the Second World War the Japanese industry operated on only a small scale for several years, but since about 1950 has more than recovered its earlier position. The industry today is somewhat differently based from that of the pre-war years. Scrap is less easily obtained and is relatively more expensive; the supply of pig-iron from the mainland of Asia is now cut off. The smelting industry has been greatly expanded,

		Steel produced	Pig-iron smelted in Japan	Imported Pig-iron	Imported iron and steel scrap
			(thousand metric tons)		
1934	..	3,844	1,728	779	1,413
1935	..	4,704	1,906	1,093	1,692
1936	..	5,223	2,008	1,095	1,497
1937	..	5,801	2,308	1,130	2,420
1938	..	6,472	2,563	1,072	1,358
1939	..	6,696	3,179	928	2,555
1940	..	6,856	3,512	855	1,391
1941	..	6,844	4,173	784	203
1942	..	7,044	4,256	882	39
1943	..	7,650	4,032	315	30
1944	..	6,729	3,157	377	222

(From *Statistical Year Book for 1957*, The Japan Iron and Steel Federation.)

and the Japanese industry is now based mainly on the import of ore and coking coal. The ore is obtained mainly from the Philippines, Malaya and India, and the coking coal from the United States. In so far as Japan no longer controls politically any of the sources of imported materials, her position is even less assured than before the war (*see* table opposite).

The Japanese steel industry is carried on at three major centres. The largest and most important is the region of Kobe and Osaka, which concentrated at first on steel-making and rolling, using imported pig-iron and scrap. There are today many small steel-making and rolling works. Hirohata is the only large integrated works in this area, which has about 30 per cent of Japan's steel capacity.

Second in importance is the area around Tokyo Bay. Plant was established here to supply the local engineering and ship-building industries. There are today one integrated iron and steel works, together with a large number of semi- or non-integrated steel and rolling works. The Tokyo region has somewhat more than a fifth of Japan's steel capacity.

The industry in northern Kyushu is third in importance. The Yawata plant was established here in 1901. It lies close to the tide-water, and uses imported ore and coke prepared by blending local with imported coal. The capacity of the Yawata works, the largest in Japan, is 2,100,000 tons of pig-iron and 2,583,400 of steel. At

	Steel production	Pig-iron production (metric tons)	Imported scrap	Imported ore
1950	.. 4,839	2,233	—	—
1951	.. 6,502	3,127	214	3,089
1952	.. 6,988	3,474	506	4,768
1953	.. 7,662	4,518	1,141	4,290
1954	.. 7,750	4,608	978	5,005
1955	.. 9,408	5,217	1,287	5,459
1956	.. 11,106	5,987	2,515	7,766
1957	.. 12,570	6,815	3,328	9,381
1960	.. 22,138	12,341	—	—
1961	.. 28,268	16,383	6,551	21,239
1962	.. 27,546	18,439	3,635	22,445
1963	.. 31,501	20,434	4,591	26,268
1964	.. 39,799	24,450	5,094	31,236
1965	.. 41,161	28,160	3,407	39,018

(From *Statistical Year Book for 1957*, The Japan Iron and Steel Federation. *Statistical Survey of the Economy of Japan, 1964*, Ministry of Foreign Affairs, 1965)

Kokura, a short distance east of Yawata, is a small integrated plant.

Other centres are near Nagoya, in northern Honshu and southern Hokkaido. They are relatively far from the industrial markets in Japan itself, but, alone amongst Japanese works, are able to use mainly domestic materials. The Kamaishi works smelt the local iron-sands, supplemented by imported ores. The Muroran works in Hokkaido smelt the local limonite with fuel from a nearby coal-field.

The table on this page shows that the Japanese steel industry has by now very greatly exceeded even its wartime production. The larger works are being modernized, and it seems probable that some of the smaller and less efficient will be closed. It seems also that fuller use is being made of existing capacity. A country such as Japan, which has to import virtually all its steel-making materials generally from a considerable distance, can sell its products to the world market only if the iron and steel industry is organized on the most efficient scale.

AUSTRALIA

Like Japan, Australia can fairly claim to be included amongst the industrialized countries. About a third of its gainfully employed population is engaged in industry and its standard of living is high. Reserves of iron-ore are large; the supply of coking coal is at least adequate for a smelting industry, and it cannot be doubted that the domestic market is big enough to justify a large-scale, modern industry.[20] The remoteness of Australia from other industrialized countries makes the import of steel goods relatively expensive. For the same reason, Australia cannot look for any large export market for her products except in New Zealand.

The iron-ore resources of Australia are concentrated in the iron ranges of South Australia and on the north-west coast (page 45–46). These are all close to the sea, which greatly facilitates transport to the smelting works. The Newcastle coal-field of New South Wales is not large in total reserves, but it does contain a high proportion of coking coal.

The Australian iron industry was established during the nineteenth century, but it was not until the Broken Hill Proprietary Company founded the Newcastle (N.S.W.) works in 1915 that it achieved any degree of importance.[21] Demand and output grew during the years of the First World War, and a post-war collapse was prevented by high tariff protection for the industry. In 1928 the Port Kembla plant was opened by the Australian Iron and Steel Company. It seems, however, that the Australian industry was over-expanded. It was a relatively high-cost producer, and during the Depression years the Newcastle plant was closed. But the industry revived in the later 1930's and, under the stimulus of re-armament, was considerably extended. A smelting works was built in 1941 at Whyalla, on the coast of South Australia, less than forty miles from the iron ranges themselves. At present the ocean-going ore-carriers, which take the ore from South Australia to New South Wales, return with cargoes of metallurgical fuel for the Whyalla furnaces. Whyalla exports pig-iron to other parts of Australia, but does not make steel.

Since the Second World War the Australian iron and steel industry has grown rapidly, and production of pig-iron in 1965 stood at 4,356,000 tons, and of steel at 5,500,000. The import of steel goods, which had formerly been large, has now been reduced to inconsiderable proportions. It consists mainly of pig-iron for the steel works and of certain types of steel goods that Australia does not yet produce in quantity.

CONCLUSION

In the last five chapters of this book the iron and steel industries of individual countries have been briefly discussed. The accident of great mineral wealth, combined with technological progress in the eighteenth and nineteenth centuries, has allowed some of these to establish industries which are large in terms of output, adequately capitalized, and capable of producing steel at a relatively low price. This group of countries includes, in addition to the United Kingdom, Sweden, France, Belgium, Luxembourg, Germany and the United States.

A second group of countries is made up of those which, though adequately endowed with resources, were delayed for historical reasons in developing an iron and steel industry. To this group belong Canada, Mexico, the Union of South Africa, Spain, Turkey, India, China and Australia. And to these should probably be added the Soviet Union. In varying degrees these countries are still underdeveloped. The railway net is by no means as extensive as is desirable. The domestic market is not large enough in every instance. Some of these countries are still technically somewhat behind the level of the first category of countries, and are still relatively expensive producers.

A third group is made up of a few industrialized nations which have the misfortune not to have been well endowed by nature. But, by technical skill, careful management and, in certain instances, by concentration on highly specialized goods, they have built up a prosperous and profitable industry. Switzerland, Italy, Poland, Norway and Japan belong to this group.

The last category is made up of newcomers to this field. They labour under immense difficulties of shortage of capital, lack of a *cadre* of technicians, and absence of one or of both the essential raw materials. While Brazil and Chile may be expected to make steel efficiently and cheaply as soon as their early problems are overcome, there are other countries whose steel industry was created by pride and politics and is maintained by tariff-protection and subvention. It may fairly be asked whether in such countries the public welfare would not be benefited by investment in some other direction.

World production of steel has been rising sharply, and this rate of increase is not expected to flatten off in the near future. Already, in 1965, world production of steel was about 443,300,000 tons, about three times the peak output of the Second World War. Cheap steel is one of the most urgent necessities of almost all branches of modern industry, and is thus essential for high standards of living. Erecting a steel-mill in Indonesia is more likely to increase the local price of constructional steel than to make steel goods more easily available to the Indonesian. It is the large, efficient, well-located mill, such as the several continuous strip mills that have been mentioned in this book, that is able to produce most cheaply. A modern, integrated plant represents so great a capital investment, and so large a proportion of the national income of a small country, that its location needs to be chosen with the utmost care. It is doubtful whether this can be done within the limits of a single country, unless it be the U.S.A. or U.S.S.R. In Western Europe the planning of new plant transcends national boundaries. It would be unfortunate if the brash, new nationalism of some countries were to reverse this very desirable trend.

[1] *European Steel Trends in the Setting of the World Market*, Steel Division, E.C.E., Geneva, 1949.
[2] *Steel Production and Consumption Trends in Europe and the World*, mimeographed report of Steel Section, E.C.E., Geneva, 1952, and *The European Steel Industry and the Wide-Strip Mill*, Industry Division, E.C.E., Geneva, 1953.

[3] *The European Steel Industry and the Wide-Strip Mill*, Geneva, 1953, 3–14.

[4] *World Iron Ore Resources and their Utilization*, United Nations, 1950, 49–60.

[5] *A Study of the Iron and Steel Industry in Latin America*, two volumes, United Nations, Department of Economic Affairs, New York, 1954; the first volume is a careful appraisal of the resources and potentialities of the area; the second consists of technical papers on appropriate industrial processes.

[6] Robert G. Long, 'Volta Redonda: Symbol of Maturity in Industrial Progress of Brazil', *Economic Geography*, XXIV, 1948, 149–54; 'Development Plans in Latin America', *Monthly Statistical Bulletin, British Iron and Steel Federation*, XXVII, September, 1952.

[7] J. P. Cole, 'Colombia's First Integrated Steelworks', *Geography*, XLIII, 1958, 56–7.

[8] *World Iron Ore Resources and their Utilization*, United Nations, 1950, 44–5.

[9] Raymond Furon, *Les ressources minerales de l'Afrique*, Paris, 1944.

[10] Peter Scott, 'The Iron and Steel Industry of South Africa', *Geography*, XXXVI, 137–49.

[11] *The Times*, 14 November, 1957; see also *New York Times*, 8 April, 1956.

[12] Pradyumna P. Karan, 'Iron Mining Industry in Sighbhum-Mayurbhanj Region of India', *Economic Geography*, XXXIII, 1957, 349–60.

[13] On coal and ore reserves in Asia see *Coal and Iron Ore Resources of the ECAFE Countries*, United Nations, E/CN.II/I & T/4.

[14] John E. Brush, 'The Iron and Steel Industry in India', *Geographical Review*, XLII, 1952, 37–55; *New York Times*, 21 November, 1957; 'The Indian Steel Industry', *Monthly Statistical Bulletin, British Iron and Steel Federation*, XXX, 1955; *Far East Iron and Steel Trade Reports*, October, 1957.

[15] *International Iron and Steel*, United States Department of Commerce, August, 1956, 12–13.

[16] *New York Times*, 8 September, 1957.

[17] I am indebted to Professor Thomas F. Barton, of Indiana University, for this information.

[18] Muzaffer ErSelçuk, 'Iron and Steel Industry in Japan', *Economic Geography*, XXIII, 1947, 105–29.

[19] *Survey of World Iron Ore Resources*, United Nations, 1950, 328–32.

[20] Clifford M. Zierer, 'The Australian Iron and Steel Industry as a Functional Unit', *Geographical Review*, XXX, 1940, 649–59; Neville R. Wills, *Economic Development of the Australian Iron and Steel Industry*, Melbourne, 1952.

[21] N. M. Windett, *Australia as Producer and Trader*, Oxford, 1933, 151 ff. See also Kenneth B. Cumberland, *Southwest Pacific*, London, 1956, 106–9.

INDEX